AF226881

GALVANISING
THE GEEKS

ZOE CUNNINGHAM

GALVANISING THE GEEKS

TIPS FOR MANAGING TECHNICAL PEOPLE

DISCLAIMER
These words are my own, and do not necessarily represent the views of organisations that I have worked for or am affiliated with.

First published 2014
By Thinksmart Publishing

Revised edition 2020
By Mind Engagement Publishing
www.mindengagement.com

All text and diagrams by Zoe Cunningham
Edited by Sean Williams and Richard Smyth
Design by Rosamund Saunders

ISBN 978-1-8382324-0-5

ALL RIGHTS RESERVED.
This book contains material protected under International and Federal Copyright Laws and Treaties. Any unauthorized reprint or use of this material is prohibited. No part of this book may be reproduced or transmitted in any form or by any means, electronic or mechanical, including photocopying, recording, or by any information storage and retrieval system without express written permission from the author/publisher.

To my wonderful husband Sean, who supports me in everything I do and without whom this book would never have been started.

Contents

Introduction

Managing technical people is one of the most significant challenges in the modern workplace. No other internal department is the source of so much frustration, and only technical suppliers seem to be able to deliver something entirely different to what was agreed.

Technical people often seem to be a breed apart – do they even speak the same language as the rest of us? And yet good management of your technical resources is key to avoiding the pitfalls of late deliveries that don't meet user need.

In my experience, gained through managing technical people at many different levels, there are times when traditional management theory works well with technical people and times when you need a new rule-book. In this book I have tried to set down some practical tips and examples to cover both types of management technique.

Often, the CIO and their team are sidelined and avoided as much as possible. This does not lead to better IT delivery! In this book, I work to bridge the gap between the ICT/IS/Digital department and the rest of the organisation.

Why this book?

During my career I have developed my skills with a particular way of learning. Instead of spending time at training courses, I have broadened my experience in a wide range of areas by learning on the job. I was sceptical about this approach when I started work, but I've since come to really value it. Even if you attend a residential training course, the real learning comes when you apply in your work environment the ideas that you have learned on the course and find out whether and how they can work for you. It is very rare that you can learn 'pure' knowledge that doesn't require some practical application.

A full-day course usually aims to teach about three key ideas. You can find out more ideas more quickly by reading tips on the internet or in books – and, if you read these in amongst your working day, you can start applying them immediately. The quicker you can find out and implement new ideas, the quicker you can learn.

Following that paradigm, I've put together a list of hints and tips that I've found useful in my career, first managing a team as a project manager, then managing individuals as a career mentor, coaching project managers as a project governor and, finally, leading a senior team as Managing Director.

Nowadays there are many excellent books on all aspects of management, and lots of the skills discussed are also applicable in the technology sector. You can find a list of resources that I've found particularly useful at the end of this book.

How to use this book

This book is designed so that you can use it in whatever way suits you best. I prefer to read straight through a book, so I've arranged the contents in an order that should make sense, but if you prefer to dip and out, that's also fine – even within each chapter, each sub-section should be self-contained.

Who is this book for?

This book is aimed at anyone who manages technical people. Some tips may be more appropriate for project managers on the ground and some more for CIOs or CEOs who are setting the tone of their whole organisation. Hopefully, most of the suggestions will be useful to you at some point in your career. If you are a senior manager you may be able to pass some of the delivery tips down to your team; if you are a project manager you can start practicing leadership skills to the benefit of your future career.

It starts with you

If you are leading your organisation or department, you will already be aware that the department's culture, ethos and training environment start with you. It's *your* behaviour that will determine how tasks are approached, right down to the delivery level.

If you are not at the head of the hierarchy, this may be less obvious – but the first place to start is still with yourself. If you don't know where you're going and how you're going to get there, you will find it harder to convince others and take them with you, whether they are your direct reports, your own manager and directors of your company, or people who aren't in your management line at all but on whose co-operation your success depends.

Your own behaviour is the one thing that you have direct control over. You can influence other people, you can persuade them, you can motivate them with carrots (usually financial) and sticks (shouting and ultimately firing) – but you cannot control them as you can control yourself. If you want your manager to adopt better people skills, you will need to get their buy-in, find a way to make it easy for them, and make sure it stays at the top of their priority list. If you want to adopt better people skills yourself, you can (within reason) simply decide to do so.

Taking the time to build your own skillset is an absolute pre-requisite to helping your team to improve theirs.

This chapter covers the following ideas.

- **People management takes the most time**
  Don't forget to make space for the most important activity you will undertake.

- **Respect your team**

  Your techies deserve respect and will do the job better if they get it.

- **Learn how to be humble**

  If you recruit superstar coders, you'll need a way to deal with superstar self-confidence, and one-upmanship will be counter-productive.

- **Don't get caught up in the detail**

  Don't think like a coder if you're working as a manager.

- **Don't develop a culture of saying yes**

  This may surprise some software firms, but you'd be surprised how often 'no' is the right answer.

- **Be the energy**

  You're the person who needs to give people a reason to come into work each day.

- **Make sure you get credit**

  Visibility is the most important factor in determining how you are perceived by others.

- **Know your strengths and weaknesses**

  If you can't help yourself it will be harder to help those around you.

People management takes the most time

Almost every manager started their career as a 'doer' – including me. When I worked as a software developer, with clear tasks to achieve that appeared to compose the majority of any delivery, it wasn't always clear to me what managers added. Apart from some tangible, visible tasks, such as liaising with the client or setting up a build server, the process of management appeared a little mysterious.

As a manager you have two things to manage: people, and process. The process-management tasks are usually the most visible. They're the ones that clearly take time: writing a report to the business, creating a risk log, phoning up third parties to find out why their interface isn't working, and so on. If you are asked to account for your time and you explain that you have been writing a report, it will be easily understood what you have been doing.

On the other hand, it can often be assumed that managing a team of people takes literally no time at all. You simply work out what everyone needs to do, tell them, give them a quick motivational speech, and you're done.

If only it were that simple! A good manager will, on the contrary, spend almost all of their time on people management. The fact that this is where most of the work is needed is testified to by the number of meetings that managers attend. Meetings are essentially a way of sharing information between people. Another example that illustrates this is the concept of middle management. Why does middle management exist? Why is there a limit to the number of people who can report into a single individual? I can track a whole company's process checklists in a single spreadsheet. I can't directly manage the whole company.

You need to have enough time set aside to engage with the individuals who report to you. A good framework to use is to think of people management as a three-step process: set goals and give a clear brief, monitor progress, and give detailed feedback frequently (including positive feedback). You can read more about this is approach in the *How to manage* sections (p62 onwards).

Exceptional, experienced individuals may be able to set their own goals, report back to you frequently so that they are bearing the brunt of the monitoring work, and even analyse their own performance. But even these individuals need you to spend a certain amount of time with them. You need to be inputting into and verifying their goals. You need to be carefully reviewing their status reports, and you need to be giving them feedback on how they are doing. Getting your team to this level should be one of your training goals – once you have individuals who can self-manage in this way, management becomes a very pleasant job.

At the other end of the scale, someone who is new to the team, perhaps a recent entrant with no workplace experience, might require a tremendous amount of support. Before you can even start on the first briefing step, you will need to build rapport so that you can communicate and feedback more smoothly. Briefing will take longer, and you will need to spend extra time verifying that your brief has been understood. Lack of a clear initial briefing is the single most common reason for underperformance. More monitoring will be required, and, until you have trained your managee to take the action upon themselves to report to you at intervals, you will need to spend time actively acquiring updates on their progress. Feedback to more junior members must also be more detailed and more copious; it is more important to follow the 5:1 rule to avoid dis-incentivisation (the 5:1 refers to the ratio between positive and negative feedback – you should give five pieces of positive feedback for every piece of constructive criticism).

Despite being the most important task, people management seems to easily drop off the bottom of busy managers' task lists. I think this is because, while people management is an important task, it seldom seems urgent. If you don't spend time managing your people, the difference may be imperceptible – at least to begin with. Over time, however, the relationship will drift away, with views and aims becoming increasingly misaligned. Once you and your team have substantially different ideas about how best to do the job, or even about what job it is that needs doing, serious problems will ensue. And to sort out these problems will take, yep, even more time.

Your job as a manager is to ensure team cohesion, with everyone pulling in the same direction. You can't achieve this if you don't put the time in.

Respect your team

In many cultures, it is accepted that those in charge will lay down rules for those whom they manage. Even if they avoid the trap of micromanaging, they feel that they can set unilateral goals. This can work well in many departments, but it won't work for techies.

Someone who chooses a technical path is someone who is inquisitive about their environment. For a technical person, every fact about the world can be questioned with a corresponding 'why?' Once someone has lived through years and years with this attitude, they get to see the benefits. They realise all the things that they have learnt by taking an interest. They appreciate the decisions that they would have got wrong if they hadn't been inquisitive.

So the first thing that will happen when you tell a developer to do something is that they will stop and think. They will consider whether it is a good idea or not. If you have a healthy culture, they will want to ask you questions about the task, about the aims and about why doing this particular thing will achieve those aims. If you don't encourage questioning then they might not ask you explicitly – but they will still consider the idea, and make up their own mind about whether what they have been asked to do would be useful. If you don't have a culture of questioning, then you probably don't have a culture of open defiance. But open defiance is not the only way to avoid doing something that you've been asked to do. If someone has not bought into completing a task, then they are much more likely to 'forget' to do it or to be 'too busy' to do it.

On the other hand, if you give a techie a good reason for doing something, the picture will be completely different. Because communication is so difficult in general, you will do even better if you let your developer challenge your assumptions: you may think that you have given a clear and compelling case; they may think that you are just fobbing them off and you have a nefarious undeclared reason hiding underneath. That's leaving aside the hidden advantage that you may be challenged in a way that leads to a better outcome or a new way of achieving the same ends. By having a discussion with a smart technical person, you might end up getting them to do some of your thinking for you.

A good example of this is reporting. Most people find reporting tedious. If you've only ever been on one side of the reporting divide, you don't have as good an understanding of just how useful good reporting can be. If you can ask with clear reasons for the exact information you need and no more, you have a much better chance of getting exactly that; a long and pointless status format is much more likely to be avoided or rushed.

Learn how to be humble

Big egos abound in technical departments. Developers are smart and they know it (sometimes they know it even when they're not quite as smart as they think they are). As most CIOs and technical leaders were once developers themselves, this arrogance tends to permeate through the whole department. A study on LinkedIn found that 90% of CIOs think that they're leaders in their field.

It's not easy to react well to arrogance. At best, it's extremely irritating; at worst, it's insufferable. Arrogant attitudes often have a counter-productive effect: if someone comes across as valuing their own opinion regardless of what anyone else thinks, it's very tempting to write them off.

Firstly, it's important to recognise any such traits in yourself. It can be easy to think that because you're smart and your customers hire you because you're smart, that you always know best. The trick is to strike a balance: on the one hand, it's important to listen closely to what your clients want and engage in dialogue to make sure that you understand the finer points before jumping in with objections or contradictions (see *Be nice to the customer*, p67); on the other, you need to make sure that you *do* challenge them, and don't just agree to what they say because they're the ones paying the bill (see *Don't develop a culture of saying yes*, p21).

As a leader of a department or team, it's even more important that you don't let arrogance cause you to ride roughshod over your team. If you peremptorily issue orders, quiet members of staff will not have the nerve to speak up – and your top techies, who may have a healthy dose of arrogance themselves, will be incensed. You may feel that you need to assert your authority and to make it clear who's in charge. Personally, I don't think that this is going to generate the kind of atmosphere where you get the best work from everyone. You need to be working *with* the superstars that you worked so hard to recruit, rather than fighting against them. You need to make sure that you don't fall foul of the tendency to let your pride blind you to the right solution, just because someone's being unbearably smug about having worked it out. This may be as simple as taking a few deep breaths, or asking for a simple written summary that you can assess later.

Finally, you need to find a way to deal with members of your team who are letting their arrogance affect other team members. Your first weapon is being open and upfront about what the issue is. Smart people, even if they are also arrogant, should understand when you have a legitimate complaint, and adjust their behaviour accordingly (provided that you give clear advice to help them – see my tips on giving feedback, p108). Make it clear what is and isn't acceptable.

Link being able to motivate a team with leadership positions. I was very impressed recently when one of our most technical developers asked for a course on soft skills for tech leads because he 'wasn't very good at those'. Don't attempt to deal with arrogant people by trying to take them down a peg or two; it feels like the obvious solution, but misses the underlying causes of arrogance. Arrogant people are not super-confident. If you are super-confident you don't need to tell everyone about it; you just sit in the corner being quietly happy with yourself. Undermining people who are not so confident will upset them and mean that they won't take the messages that you're giving them as well as you would like. Instead, praise them for things that they do well (you can never overdo positive feedback – see p113) and be clear about what they need to change and why.

And don't forget to try and interview for excessive arrogance in order to prevent problems before they happen. I once interviewed one particularly arrogant candidate, who was clearly smart but was also a little bit unpleasant with it. On the final interview question he refused to take my hints regarding the best way to go about the problem. Usually I would insist that he do the exercise the simpler way, to give him the best chance, but I'd lost my patience with him – so I let him do it the long way. He got himself into a muddle. He didn't pass the interview.

Don't get caught up in the detail

I was not the best software developer. This is my key strength as a manager. I'm not being falsely modest here, but part of the reason that I can admit to it so freely (now) is that I've learnt that two key management skills have been much easier for me as a result of this deficiency: first, having sympathy and understanding for other developers who are under-performing, and, second, not getting caught up in the detail.

If you, unlike me, were a top techie (and it's not unlikely because in most organisations this is how promotion works), the biggest pitfall that you're going to encounter is feeling the need to do it all yourself. This is where I have the advantage. I don't have that luxury.

One reason why you will be tempted down this track is that you can do the job better than anyone else. You probably have years of experience at doing the job in question. It may well be quicker for you to do the job than to train someone else to do it – or, rather, it may be quicker the first time. If a job needs to be done time and time again (and most do) it will soon be cost-effective for you to have trained someone else. Don't forget that, if you can train general learning skills (how to use Google is a good start), techniques learnt on one task can be cross-applicable to others, even if a particular task originally seemed like a one-off. When delegating, you need to remember that delegation is almost always from someone who has a greater ability to someone who has a lesser ability. This is a fact about the world with which you need to come to terms. Sure enough, once your managee has reached the same skill level as you, they will be promoted to manage another team.

The reason that it is important for you to avoid the detail is that you don't have time to look at everything. You are employed to manage. It may well be that you would rather work as a developer, getting on with coding in the exciting technology of the day rather than filing reports, dealing with clients and other managers and rolling out initiatives across teams. If you want to be a developer (and are prepared to take the pay cut), then do it! If not, get on with being a manager. That's what you're paid for.

If you have responsibility across a set of teams, you need to be looking at the 'big picture' items that need to be addressed at a high level. You are a kind of aggregator across the delivery teams. If you are spending your time looking at the details of one project – even if it is to provide a critical fix – there may be bigger problems elsewhere that it was your job to foresee and sort out. You may be fiddling while Rome burns. Even worse, when you return to paying attention to pan-departmental issues, you may find that you are too closely linked to one team. Having experienced their problems more closely, you may prioritise them above others. I'm not saying that finding out what's going on in order to inform the decision that you have to make is a bad thing; just don't get caught up in the detail, and make sure you keep a balanced view.

A problem that interweaves with that of delegation is prioritisation of longer-term issues. In general the details that will grab your attention and suck you in are the ones that need to be dealt with urgently: an issue is holding up a deadline; the server is down; the build server needs to be fixed so that code can be checked in. But by getting distracted by short-term issues, you may miss the longer-term ones. Do you have a good recruitment plan? Are you retaining your staff? Are there any problems with short-term resource shortfall a few months away that you need to think about now? If you let yourself get behind by spending too much time dealing with details, you'll find it harder to make longer-term plans and may end up rushing from crisis to crisis.

Another problem you may encounter when you feel the need to delve into the detail is that the people whose job it is to worry about this feel undermined.

Don't develop a culture of saying yes

Outsourcing and offshoring are constant topics of conversation in technical departments in Europe and the US. The consensus seems to have crystallised into two basic principles: yes, offshoring is here to stay, but no, offshoring can't be used for every type of problem.

The practice of working with companies based several time-zones away has been the subject of numerous criticisms: that communication is too hard, that teams won't gel, that incentives will be misaligned. The one that I came across most when working with companies in India is one that didn't feature much in our original analysis but turned out to be an absolutely crucial stumbling block: some Indian companies had a culture of saying yes.

It doesn't sound like a big problem, does it? In fact, saying 'yes' sounds like exactly the kind of thing that you should be doing as a business. We even have pat phrases such as 'The customer is always right'. The problem is that if you say 'yes' uncritically you may end up disappointing someone later. If your answer to the question 'This piece of software is really important to me and I must have it by the end of the week – can you do it?' is 'Yes', even when you know that the answer is really 'No', you are eventually going to let your customer down.

Companies all over the world employ this 'yes now, no later' tactic in order to secure work. If you're not really committing to delivering what you promise, you can absolutely blow the competition out of the water. But for a sustainable business, where you retain your clients and win more work from them in the future, this is not going to pay off.

So in this section I'm going to talk about the perils of saying 'yes', and why it's often more important to say 'no'.

There are many things that you will be asked to do as a software team. Some examples include:

- Deliver to a certain date
- Cut standards to deliver to a certain date
- Tell your team to work harder/longer/every weekend to deliver to a certain date

- Fix bugs quickly rather than spend time fixing them for good

- Deprioritise work on site resilience in favour of new features

There are two different types of request here. The first are things that you simply can't do. If the work can't be completed by a certain date (or you think there's a good chance that it can't) you can say 'yes', but you can't mean it. At some point 'yes' will turn into 'no', and everyone will be unhappy.

It's possible that you might have built a relationship with a business where they are happy to set difficult deadlines and let you miss them as long as you tried hard. This is the second kind of request: the kind that conflicts with what your professional experience tells you is the right thing to do. If there's a chance that the person to whom you're delivering software won't get it when they need it, they won't be able to promise it to anyone else. So either they end up letting people down too, or there was no need for you to promise it in the first place.

The other cases in the list above are much clearer examples of this. If you are developing a piece of code properly, you are structuring it to minimise the total cost of ownership. If you cut standards now in order to ship a new feature, there will be an associated cost later on. This might be OK if there's going to be time later on to catch up on the missing code later – but will there really be?

Decisions such as fixing bugs for good versus patching them over with sticking-plaster, or ensuring that backups are in place in case of server failure, can sometimes seem more straightforward. There is a simple choice of priorities, and the business can choose, right? What you might be forgetting is that there is an unwritten contract that requires you, as the person in charge of the technology, to prevent future issues and cover any risks. You are employed to take the responsibility on yourself to look after the technical side of things

– even though it may not seem like it when you're being asked quite strongly to make a particular choice. In order to be able to deliver on this responsibility, you need to know how to say 'no'.

I'm not saying that you shouldn't look for creative solutions. Finding a way to say 'yes' is a great skill, but you can only get to this point if you are first able to say 'no'. Answering 'yes, if...' is often tantamount to 'no'. You need to take a stand on what is and isn't possible in order to creatively engage with what you're being asked to achieve. In life, the only good deals that you can make are the ones that you are prepared to walk away from.

Be the energy

It's fair to say that 'energy' isn't a word that's often used in software team management books. Technical people prefer concrete terms that are easy to define and possible to document. But there are some key management techniques that simply can't be treated in this way.

I think that most people who lead technical departments need no help whatsoever with planning and scheduling, with the management of the inanimate parts of the process – but dealing with people is different. I've come across a great phrase that sums this up: 'With processes you need to be efficient; with people you need to be effective.' It's no use being efficient with people if you're not effective as well. If you cut your management time down to 30 minutes each week but don't achieve the leadership and communication goals that are your remit, you have failed; you have tried to reduce your objective-based people management to a process.

Thinking in this way, I find 'energy' to be a useful term for characterising what I need to do as a manager. It is not a scientific or rigorous term; by saying 'energy' I will convey a slightly different message to each reader. So be it.

Another fact that you need to accept in order to understand why I want to talk about energy is that there is more than one way of looking at the same situation. One person's opportunity is another person's obligation. It is possible to be doing the exact same task on two different days and be overjoyed by it one day and desolate the next.

Working towards your short-term delivery goals and long-term career goals takes energy. If you want to achieve something, you often have to battle with the fact that you can't achieve it right this minute. If you want to be playing with the latest tech, you need to steel yourself to work through the task you need to complete first on the legacy codebase. If you want to become an expert, you need to work hard as a trainee.

I didn't understand what a difference being around people with lots of energy made to me personally until I moved into a business development role. Business development is all about working with people from different organisations, which often have very different cultures. Software organisations are usually very efficient. We're often horrified by the number of all-day, whole-team meetings that other organisations hold without even breaking a sweat. *All* day? *Whole* team? Do you know how much that costs?

Sometimes development agencies or technical teams aren't the highest energy of environments. When I started meeting up with individuals from client organisations I was struck by how enthusiastic they were about everything: 'We could publish everything on tablets!' – 'We could create an online world where people can explore our products!'…. It wasn't the nature of the ideas that they were coming up with; it was their approach to each idea. Why shouldn't we try to bring that kind of enthusiasm and energy to everything we do? Sure, it's tiring, but it's much more effective and rewarding. In a similar vein to the maxim that, 'if a thing's worth doing, it's worth doing well', I think that, 'if a thing's worth doing, it's worth doing enthusiastically' – the 'well' will follow.

That's my recommendation at a personal level. As a leader, though, your responsibility is for the team. I believe that you have a personal responsibility to come in to work every day full of energy. Energy is contagious: if you want a happy and enthusiastic team, the single best way to achieve this is to emanate a happy positivity.

Make sure you get credit

In some professions, the nature of the job means that you get very used to positioning yourself to take credit for things that you've done well. If you're a marketer, it's a case of marketing yourself; if you're a public speaker or a salesperson, you won't be retiring when it comes to explaining things verbally. If, on the other hand, you have worked in a technical background, you'll probably be used to letting your code speak for itself. Even worse, if you're a non-technical leader of a technical department, you may not feel associated with the success – the credit should go to the people who wrote the code, surely?

Both of these attitudes are wrong. And it's your responsibility to make sure that what your department does is recognised.

The first step is to make sure that you recognise the people who are working for you, and that you spread recognition of them further afield. If you have great technical people without whom success would not have been forthcoming, make sure people know about it. If everyone has been working their arses off to deliver, credit should be given for this.

The next step is to realise that it is exactly these successes that are *your* successes. The fact that your team were the ones who made it happen does not diminish the part that you had to play – it *is* the part that you had to play! If you've recruited the best people and have a culture that means that they stay with you and can perform to the best of their abilities, you are running a successful department! At your level you are judged on results, and your team is the means by which you achieve them.

Self-promotion and publicity often don't come easily to CIOs and CTOs, so here are some simple tips. Remember that, as with everything, these will get easier to follow the more you practise.

1. Make sure that your successes are included in publications that circulate around the company. Make friends with the Communications department; remember that they may not realise how challenging or critical your work is – you may need to do some work to explain this, and this is part of your job!

2. Write your own newsletter to be sent round at the same time, or as an inset, allowing you to explain exactly what the most important technical developments are.

3. Make a point of going to talk to heads of other departments; make sure that they know what you're doing – and how you can help them.

4. Gain recognition for your department outside of your organisation. Do you have a great development methodology? Are you adopting cutting-edge languages in unique ways?

5. Share your ideas and achievements more widely by writing a blog (you can see the Softwire blog at https://www.softwire. com/insights/).

6. Talk at conventions. Influential tech leaders can reach a large audience by talking at conferences – describing their early adoption of Scala, for example.

7. Build partnerships with other software organisations; teach and learn.

Putting all these ideas into practice won't always be easy – but it's certainly worth the effort.

Know your strengths and weaknesses

Unbiased self-assessment can be extremely helpful in working out where you want to get to. Knowing your weaknesses can stop you haring off down a path that isn't for you, or help you work out how to focus your training efforts on the things that are key to securing your dream job. But it's important to focus on your strengths too – unless you know what they are, you won't be able to maximise and build on them. If you're great at something and don't realise it, you're probably not making the most of your talent.

Working this out for yourself can be hard, particularly if you're in a position of authority. It will sometimes feel like it's in the best interests of others to bite their tongues rather than give you their unbiased opinion (even if this isn't the case). Try to get input from trusted colleagues – and even from outside the workplace, if you feel that your behaviour there is sufficiently similar to the workplace 'you' (it isn't always!). Then use this opinion to gauge your performance objectively. Try to become the person who can tell you what you're doing well and not well. It will be easier once you appreciate that, as with everyone else, there are some things that you're good at and others that you're less good at. If you can remain undismayed by your weaknesses and modest about your strengths, you'll have a better chance of seeing them for what they really are.

As with everything when you're a senior manager, you need to be able to help your team with this process of assessment too, so that they can enjoy the same benefits.

Firstly, align assessment (see p141) by making sure you're clear with them about how you see their strengths and weaknesses. Encourage them to seek out other opinions, and to build on all the opinions that they receive in order to create a balanced overview. Show them how to work around weaknesses, build up weak but critical areas, and utilise strengths. Find and suggest tasks that will help them with one or all of these. If everyone is honestly and openly examining what they are good at, it will be easier for individuals to join in.

Leadership

Leadership is the one skill that is never taught. Despite the (long overdue) acceptance of the fact that skills, expertise and even genius are the product of hours and hours of hard work and learning, leadership is often still considered to be a gift from the gods. Some people, we're told, are simply born to be leaders.

Some people start in a leadership role as better leaders, perhaps because of 'natural ability' or perhaps simply because their childhood learning has encouraged the development of these particular skills. Whatever the reason, we can all recognise a good leader when we see one, and yes, those people will have an advantage starting out in a leadership role. It's important to know that if you're not a 'natural' leader you can start from scratch and learn the skills that that you need to do your job. In fact, often having learned the skills you will be better able to hone and appreciate them than people who didn't have to try as hard.

This chapter covers the following ideas.

- **Instil passion**
  People who love their work deliver better results.

- **Lead from the front**
  Don't ask your team to do things that you wouldn't be prepared to do yourself – model the values that you're trying to spread.

- **Take pride in technical excellence**
  We are all in a business with tight time-pressures that can cause us to skimp on quality – don't do it!

- **Share the knowledge**
  Keeping everyone informed on key business ideas will help them to do their job better.

- **Be inclusive**
  Leather desk-chairs and private offices are *so* last century…

Instil passion

It's often assumed that work is the opposite of leisure. Economists draw graphs to show how much you need to pay people in order to get them to relinquish valuable leisure time for toilsome work.

Yet this doesn't always hold true – and it certainly doesn't for many software professionals. Software professionals often finish work, go home, and – wait for it – *do more software development*. They code mobile phone games. They contribute to open-source software projects. They write scripts to make their home PCs function more efficiently.

At least, the good ones do. Malcolm Gladwell's book *Outliers* puts forward the theory that to be an expert in any field you need to put in 10,000 hours of effortful practice. If a software developer is going home and putting in another four hours on top of their day job, they'll be increasing their skill levels 50 per cent faster than their coworkers (of course, it does need to be *effortful* practice: see p94 on goal setting.)

So if a developer is passionate about their job it will result in benefits for them and for you. But you can't control what a developer does when they're at home; the chances are that saying 'Go home and have some fun developing something!' isn't going to work.

You're going to have to start by helping to develop their passion for their work *while* they're at work. Find out what attracted them to the job in the first place. What were they looking to achieve? What did they think would be fun? The obvious solution is to match people with projects that they're going to enjoy, but that's not always possible. The less obvious solution is to find interesting challenges within the project that they're working on. Are there interesting integration points? Can they try out new tools or technologies? Can they collaborate or pair-program with a more passionate member of the team?

Lead from the front

It's easy to think that, now you're a manager, your job consists of doing something completely different to those beneath you – that you have nothing in common anymore.

It's not easy to follow a leader who thinks like that. Indeed, why should you? If they are doing a completely different job to you, why should they know anything about what you do?

However big your department, you want it to function as a team. You need to be sharing the good times and the painful times. If a team is working late, you need to be part of that. One of the most rewarding weeks of work that I did was standing in for a senior project manager while he was on holiday, mentoring a junior project manager in charge of a team whose project had very challenging deadlines. It was actually a great time to be on the project, because they had just turned the corner and, while they weren't yet in the clear, the burn-up (progress) chart was heading in the right direction. Being there when they worked late and coming in for an early team breakfast was (an enjoyable) part of the job.

You also need to make sure that you are around enough to be able to be a part of the larger team. Of course, you will have many demands on your time that take you out of the office, or into meetings – so make some time in your diary to be visibly part of what's going on. You never know, you might learn something too.

Take pride in technical excellence

Life is a series of compromises. You want to deliver the best you can and to serve the business in the best way; even so, you will often need to bargain with others who see the world differently, or have gatekeeper powers and oppose your plans.

From a developer's point of view, customers (external or elsewhere within the business) are always forcing them into compromises. They want another week to finish a feature properly – the customer wants to ship now; they want to use a new technology as an experiment – the customer wants to use a well-established standard tech to ensure that they can maintain the codebase going forwards.

Pragmatism has many benefits. Indeed, it's not possible to run a business without it. Sometimes approaching the code in a certain way won't be the right decision – but everyone should be aiming for realism, not perfection. This is the right way to approach decisions at a macro level.

At a more detailed level, however, 'pragmatism' is often simply a euphemism for sloppy work. While sometimes it makes sense to leave large features that can be implemented later to a future sprint, or to choose a simpler implementation now that may have to be rewritten later, it's never correct, in my view, to 'save time' by writing poorly documented, poorly structured or poorly tested code.

The first reason for this is one that you'll doubtless be aware of, but that bears repeated mention: when costing a piece of software, the most useful cost to look at is the total cost of ownership (TCO). If you consider merely the build cost, or the cost of the first phase, the parameters that you are working to will give you the wrong outcome.

Of course there will be circumstances where you don't know whether there will be a Phase 2 or not, but this doesn't negate the basic principle. If you don't know what time period and feature-set the total cost will be calculated across, you need to take the decisions that minimise TCO across all possible outcomes, allowing for the probability of each (and possibly some other factors too: if the business fails, you will have wanted to minimise the costs, for sure, whereas if you are successful you may have more cash available to improve the software). In reality, rather than this being the exception, you are almost always in this position.

Total cost of ownership is an important concept, precisely because we often forget to factor in the support costs of the future, or we make unrealistic optimistic assumptions. Thoughtworks, the international software provider, estimates that 86-94 per cent of software cost is incurred post-release. So, if you spend £100,000 developing a system, there is another £900,000 of costs to come over the remainder of the life of the software.

Good coding practices cost money *now*. Adding in automated unit testing, spending time on code structure and researching best practice takes time and costs money. But considered over the whole lifetime of the software, these practices *save* money; they reduce the total cost of ownership. This is of course why they are such hot topics, and why software teams everywhere are focussing more on writing code that is excellent at the detail level.

This is an unarguable benefit. But I think there's an even stronger one. I think the most dangerous outcome of writing sloppy code in haste to meet a single deadline is that what is done once will get done again. If you accept deadlines that force you to compromise code quality, these compromises will slowly and inexorably creep into your work until they are the norm. Once you've established a culture of hastily written, imperfect code, it will take hard work to undo it.

This is a typical 'pay now or pay later' choice. Even if it isn't always expedient, the habit of 'paying now' will mean that you reap benefits long into the future.

Share the knowledge

As news outlets constantly remind us, we are living in a golden age of information. Thanks to the glorious internet, scientific papers, howto guides and other information have never been so easy to access. Being able to use Google well can increase your capability many times.

But there's still an advantage to holding information in your head – or at least holding the knowledge of how and where to find the information. More than ever, there's a need for verified, expert sources of knowledge to help you cut down your search-time. Assuming you're running a shipshape organisation that has recruited, trained and retained the best developers, you have just such a source right there on your doorstep. But do they know it?

Developers tend by nature to be solitary workers. The difference in noise volume between techies and other teams is palpable. In a sales team, new ideas may well be shouted across the office as soon as they arise; new development ideas, on the other hand, might only be discovered during a code review.

To maximise the working of your team, you need to encourage sharing between your developers and, most importantly, from senior to junior team members. A wiki is essential as a place to store common training material and hints and tips. For time-sensitive but historically available material (which could include notifications about new wiki posts), a blog works well. *Ad hoc* group emails and presentations should be encouraged, and establishing regular presentation slots can remind your developers that you value their contributions across the company.

An effective way to share information is to encourage regular presentations from everyone in the company. Lunchtimes sessions can work well if people have trouble finding time to attend a talk during 'work hours'. Another format that is good for new speakers to try out is that of 'lightning' talks, where each speaker presents on a topic for just five minutes.

Be inclusive

You probably already have a communication plan for keeping everyone up-to-date with your strategy for the department. Communication is an essential job of those who lead; specific resources in your own or other teams will be often available to help you with this important job.

But it's easy to view communication as a one-way exercise. You and your senior team are collating information from throughout the department, deciding what needs to happen, and communicating this back. You spend a lot of time considering the pros and cons of each of your strategic decisions, including intricate details that can be too subtle to include in subsequent communiqués.

In some businesses, workers are still viewed as drones to be perfunctorily informed of their employers' decisions – much as they were in the days of the Industrial Revolution. Decisions were taken and instructions given. This will not work in a technical environment.

Your techies are smart. If you've focussed on your recruitment, they are really smart. And if you've perfected it, they are smarter than you. Smart people think about things. They will not only want to know the 'why' behind every decision; they will also formulate their own opinions. They will ferret out the complications and the special cases, and wonder why you aren't considering them. This will happen even if you *are* considering them – and even if you're communicating them, too (we've all switched off at company presentations before, right?).

Often the usual way of things will be accepted, and, even though some people are unhappy with some decisions, business will go on as usual. Every so often, though, a problem will be big enough to make important individuals unhappy to the point that they want to leave, or groups of people unhappy enough to focus their conversations on the problem, creating a significant pocket of dissatisfaction.

The best way to deal with this is to include everyone directly in the decision-making process. I'm not suggesting that you invite everyone to board meetings, but do consider setting up other forums for input, such as monthly meetings that everyone can attend once a year (and maybe the keenest people can attend more often), where any topic of conversation is up for debate.

This is a key belief of mine. It's always tempting to hold back some information. This could be because you don't want to have to spend time debating the decision (again...), or because you know that certain groups of people will be unhappy with the outcome and you're hoping that, if you don't discuss it, they won't notice that the decision has been made.

The former is laziness; the latter is gutlessness.

Debating your decisions is a key part of communication and is a part of your job that you need to embrace even if you don't relish it. Attempting to ignore the problem is sure to come back and bite you further down the line; in most cases, the group that you're hoping will forget about the new policy will know exactly what your game is! The only thing worse than someone implementing a stupid policy that makes your life difficult is someone implementing a stupid policy that makes your life difficult and pretending that they didn't.

Sharing information and encouraging input will also bring benefits that can't be realised in any other way. Aside from the positive effects on your team's happiness, which it is very much in your interests to maximize, you will discover new options that you hadn't previously considered. There are myriad ways to approach any problem, and you won't find them all on your own. By openly explaining what you have considered and what the parameters for the decision are (and why), you'll allow your techies to present you with other, often better options.

Tips for Techies

This book is aimed at two types of senior IT manager: those with a technical background, and those who have entered the role from another management function. Most management at a senior level doesn't require technical knowledge, and in fact technical managers run the risk of getting sucked into the detail (see *Don't get caught up in the detail*, p19), but still, sometimes you may be called on for help, or you may wish to explain why a task that seems impossible to one of your junior managers could be achieved with better planning. This chapter aims to give you some high-level hints and tips for tackling technical issues, focussing on the following concepts.

- **Time management**
  Someone who gets twice as much done completes a project twice as fast at half the cost. I'll offer some ideas to help you squeeze the most out of your time.

- **Realism, not perfection**
  If you don't aim for high quality you will encounter issues (see *Take pride in technical excellence*, p33), but if you allow your tech team to strive for perfection above all else your project won't get finished.

- **Do hard things first**
  Human nature leads everyone to be tempted to start down the easy road, but if you leave difficult tasks till last your schedule is likely to slip.

- **Beware large numbers**
  Breaking down tasks makes it easier for everyone to estimate them more accurately, and for hidden assumptions to be uncovered.

- **Don't develop what isn't needed**
  80% of users make use of only 20% of the functionality –
  get more bang for your buck by taking tips from the Agile
  methodology.

- **What information can be documented?**
  Don't reinvent the wheel or have people stuck without a
  handover – *write it down.*

Time management

Time is money. Sometimes people take a while to work this out – or
completely fail to take it into account – when planning. In software
teams it is not so often forgotten. If you work for a software services
company, you can't but notice the amount you could be charging
to a client for your time – or, if it's you that's charging the time,
then your client will certainly notice! This culture extends to other
software departments too, although there teams are often managed
more by deadlines – it's less 'time is money' and more 'there isn't
enough time in the day'.

In this kind of environment it's very tempting to try to control
time. To a certain extent, of course, that's exactly what you *should* be
doing. You need to control your personal time. You need to optimise
everything you do so that you can be as effective as possible. You
need to take a lot of care when scheduling meetings for large numbers of people. How much time will you be spending collectively? Is
this the best use of everybody's time?

In a senior role you don't have the luxury of squandering your time.
There are countless things that you need to do, and you need to be
ruthless in setting your priorities.

When you're working every hour you can find and fighting so hard
to fit everything in, it becomes really tempting to try to squeeze

more out of your developers. Why are they playing pool? They could be coding up that extra feature for you. Are they optimising their email-reading around their coding? Should they be working on this piece of code today, and that other tomorrow?

There's a good reason why you aren't best-placed to decide this. You are best placed to optimise your own personal time, because you know exactly what you're trying to achieve and what the competing demands on your time are. If you try to advise a random developer in your organisation, you won't have the same detailed knowledge of what they're trying to do. Sure, *someone* needs to be worrying about optimising their time, but that someone is them, not you.

It might seem that this shouldn't apply to playing pool, or fussball, or any of the million other games you make available for your developers (and if you haven't made games available yourself, the internet will more than compensate). People who are playing pool when they should be working are just lazy, right? They should be told to get on with their work. The pool table should be taken out, the fussball chopped up and dumped on a skip. And, if you want to take this to its natural conclusion, you should white-list the internet sites available.

The problem is that by doing this you're confusing hours spent at your desk with hours of productive output. The two are not equivalent. In fact, in long-hours cultures, the pressure is often off to work as efficiently as possible, because you can always make up time by working all evening. I never work as hard as when I have a mountain to stuff to do and I have to leave by 6pm because we have people coming round for dinner. In that situation I not only work hard, churning through the things with an imminent deadline, but I also work smart – re-planning and re-prioritising, working out quicker routes, and so on.

If you ban games and other diversions, you are taking some freedom away from the developers. Remember, your aim is to build a culture of empowered developers who are able to make good decisions, see the things that you don't see, and optimise their own workloads. If you start taking away their freedom, saying (implicitly, but very clearly) that they can't be trusted to make their own decisions, then your attempts to try and empower them will fail. Empowerment and trust are two sides of the same coin.

In any case, you won't necessarily find that the developers who work through the whole day without stopping for a break are the most effective. The type of brain-work that your techies are doing simply can't be kept up indefinitely. If you force people to work when they're tired, they will make mistakes. Mistakes take a lot longer to rectify than the time it takes to play a frame of pool.

So should you *force* your team to take regular breaks? Again, I'd suggest that the best person to decide this is the individual who is doing the work. There are many different theories on how you should structure your breaks. 'Pomodoro' technique, for example, suggests putting aside anything that might distract you and working straight through for 25 minutes, then taking a five-minute break (the term Pomodoro refers to a particular type of timer that can be used for this). But the same break structure won't work for everyone. The most important thing is to get people to try new things (see p175) so that they can work out for themselves what break patterns work for them.

It's true, of course, that too many breaks can lead to not enough work. But it's your job to get the most from your team, and the most effective way to do this is by letting people work to their own time patterns and judging on results – by setting goals, and measuring people against them.

Realism, not perfection

I recently re-read Robert Tressell's 1914 socialist novel *The Ragged Trousered Philanthropists*. One thing Tressell discusses in earnest is the concept of 'work for work's sake'. He offers a very clear view of the difference between rushed, unchallenging and hence unfulfilling work, and work that uses all of one's talents to do something well and produce beautiful output. In this regard, the novel reflects an idea that has been very commonly discussed in more recent management texts.

The difference between intrinsic and extrinsic motivation was first discussed in the 1970s. Extrinsic motivation is motivation provided by external factors: higher pay, for example, or a gun to the head. Intrinsic motivation, on the other hand, is driven by interest in or enjoyment of the task itself, and exists within the individual. As these two forms of motivation have been investigated more thoroughly, intrinsic motivation has come to be thought of as the more powerful; it is certainly the more enjoyable.

This fact provides an incentive for both you and your developers to do high quality work. Doing a job well is rewarding and motivating – and hence productive and profitable. You will find that your more technically proficient developers do this naturally and to a high level, thinking carefully through all options and constructing well-architected, future-proof pieces of software.

Now I am in no way advocating not doing high-quality work (quite the opposite, in fact) but there is a line beyond which this is no longer a good use of time and money. As with all such lines, it is not possible to draw it exactly, but it's clear that on one side of it lies a rough and shoddy job and on the other an over-engineered solution that will never be used to its full capacity.

Agile development methodologies add an interesting slant to this thread. Agile development advocates both 'just in time' development – i.e. not working to a complete system specification but developing the most important features first and making code extensible – and re-usable code, easy to change if required and easy to build on in the future.

I think that these are the right things to be considering, but it would be disingenuous not to acknowledge that there is a trade-off between the two. There are going to be decisions to be made about where certain resources should be stored, and the ultimate design that you are working towards is going to make a difference. Even if your codebase is super-optimised for modification there is still going to be some cost to rewriting.

Another thing to consider in the quest for perfection is search cost. The options available at the design stage are very often infinite. By thinking harder, you can always come up with a new solution (albeit with diminishing returns in terms of the probability that it's the right one). A neat phrase has been coined to describe what happens when too much time is spent thinking and planning rather than implementing: 'analysis paralysis'. If only perfection is good enough, you'll never be able to stop searching, and, rather than perfection, you'll end up with nothing at all. Better to work to a solution that is good enough and then stop. If you plan your code with an eye to changes in the future, you can always go back and implement a 'more perfect' solution later.

Do the hard things first

Every team and individual works to some kind of 'to do' list. It might be in their head, or written down in the form of stories, or stored elsewhere, but we all have one. I'd advocate that everyone has this list written down: it's going to exist in some form in any case, and by writing it down you'll be able to control it better.

Some ordering will be applied to this list. One option, which will be used in only a very few cases, is to arrange the list in the order in which things were added to it – either FIFO (first in first out) or FILO (first in last out), depending on whether you are more methodical or reactive. However, most people will be driven by their subconscious to apply some level of ordering to tasks.

One very common means of ordering is in fact the very opposite of a good order for development tasks. If your subconscious is inclined to lead you along the easy path through life, then you will find that the more simple tasks will tend to drift to the top of the pile. This may result in you feeling very efficient – you'll be burning through loads of tasks, and probably even getting ahead, which has to be good, right?

Wrong. The apparent gains you make by taking on the easy tasks first are in fact the opposite; they constitute a debt of hard tasks and risk that you're building up for later in the project. Your burn-up chart will eventually tail off as you hit the harder, less-well-estimated and riskier tasks. And the later you uncover these problems, the fewer options you have for dealing with them.

The best tasks to do first are therefore the hardest ones – the ones that fill you with dread just from looking at them – the ones that you really do not have a clue how to approach – the ones that you hope might go away if you ignore them. They won't.

This is a corollary of good risk management (covered in detail from p116). Remember that the key to dealing with risks is to get them resolved, one way or another, as soon as possible. Hard tasks are essentially risks. If you don't know how you're going to do something, how do you know that it's even possible? You certainly don't know whether it can be done within the time allocated. If something is difficult, there's a good chance that it will take a lot longer to solve than you've allowed for.

The other tasks that should be prioritised first are those that are in some way dependent on other parties. If you have developers who don't like to interact with irrational, unhelpful and unreliable third parties (and let's face it, who does?), they may be naturally inclined to leave such tasks until they have no choice but to do them. Of course, it's the unreliable nature of third parties that means that you need to leave plenty of elapsed time (if not actual working time!) to deal with whatever inconsistencies and changes you might uncover, and to work around system outages, holidays and all the other blockers at their end. How many times have you been given a third-party interface definition and tested against the test site successfully, only to find that the live site doesn't use quite the same format? How many times have you left your testing to the last minute only to find out that the test server is down? The weapon with which you can overcome these pitfalls is time. Make sure you have enough of it.

'Release tasks' deserve a special mention at this point. They're easily overlooked, for all the reasons I've just mentioned. They can be very much 'out of sight' during the development phase of a project – they are simply not part of the tasks that you're doing now; they are part of the next phase. What's more, they often hinge upon many unknowns. You might not even know whose job it is to figure out the unknowns. If you're going to have to deploy on to hardware that's owned by another department or even a different company altogether, there are all kinds of potential hurdles that you'll need to just identify before you can start planning for the rollout of your software. Some of these things you can do yourself, but some will need the involvement of the peeps who run the hardware. Whilst development is still ongoing, dates, exact user characteristics and all kinds of other essential facts can be completely unknown. And this is exactly why you need to start planning this phase nice and early.

If your team is going about its tasks in the wrong order, it will manifest itself in late or rushed deliveries, last-minute changes and a general air of panic as you approach the end of a project. Don't be tempted to leap in! Managing your team to help them achieve their objectives in such a situation must still be done in the same way: through empowerment and delegation rather than micromanagement. Insist that they try a new way, and let the results speak for themselves.

Beware large numbers

It's a sad fact that a lot of software projects fail – a little over 60 per cent, according to most estimates. If you include the number of projects that are merely late, shelved or not fit for purpose, this figure heads ever northwards.

The single most common cause of failure for software projects is misalignment between the time that the project is expected to take to develop and the amount of time that it actually takes. There's a good reason why estimating is known as 'the black art'.

Assuming that you're far enough down the line not to be estimating against tasks that bear little resemblance to those that you'll eventually undertake, there are two main ways in which your estimates can be horribly, horribly out.

Firstly, single tasks may overrun by hundreds of percent. The coarser-grained your estimate is, the more likely this is to happen. I've found that tasks that are in the order of a day or two are easiest to estimate. It is too easy to underestimate the size of a 'large' task; there are very few tasks that don't feel like they will fit inside a 20-day estimate – that is, until you break them down. Even a task of five days is likely to be concealing hidden *gotchas*, simply because, to get a five-day estimate, you don't need to know exactly what you'll be doing on each day. So Rule One of estimating is to break down large estimates into smaller ones.

The second most common cause of overrun is The Missing Task. You can have a whole spreadsheet of perfectly estimated line items, but, if there is just one task that isn't on there, your project will slip by the amount of that task. To help avoid such omissions, it can be useful to have a standard set of tasks that you always estimate (or explicitly omit) for every project – and to make sure you think carefully about what *exactly* needs to be done. Estimating in small numbers, as described above, can help in dealing with omissions that may or may not have been included as part of a line item.

It's hard for anyone to assume that there are tasks hidden within a single day estimate – but however thorough your process and however rigorous your analysis, you will still miss some tasks. It is inevitable. So you need to leave some contingency, preferably just by adding a big fat contingency task as a new line item. Without any contingency, you are guaranteed to fail.

Even if you have estimated as well as you can, there remains a third common cause of overrun that you will need to deal with throughout the project: changing requirements. Projects can be perfectly estimated and brilliantly executed, but, if they are creating the wrong things, they will be at best late and at worst canned. It is common in today's software environment to have an explicit, often non-technical, owner of the scope and output of the project. This person is called a product owner. A good product owner is worth their weight in gold. It is not an easy job. All but the most trivial of projects have several stakeholders, all with competing interests, and managing these into a single output system is an extremely difficult task.

As the technical team, you unfortunately cannot assume that there is a great product owner available to you. Having a busy or overworked product owner – or, worse, no product owner at all – is a gigantic risk to your project. And you don't have the luxury of throwing up your hands and claiming that it was someone else's problem: if the requirements are not looked after the software output will be poor, and the software output is your responsibility. If it's not up to scratch, there's no way for you to dodge the blame.

You will need a strategy for taking on the product owner role. If you have qualified (and brave) members of your team available, you can create a 'shadow' product owner. If not, you'll need every developer to do their part in keeping the scope of the project under control.

You'll need to work on fine-tuning your communication skills so you can ensure that everyone is agreed on what's been decided. If you receive contradictory instructions, you need to identify and resolve them. Most important of all, you will need to stand up to the business (Don't develop a culture of saying yes, p21).

A particularly sharp incentive for getting estimating right is to work on a fixed price basis. Then if you fail in your estimating, your developers go hungry (or least bonus-less). This doesn't work so well if your client is part of the same organisation and it may be tempting to allow yourselves to get away with being poor at estimating; there are almost always excuses that can be invoked if a specific project fails. But each time you do this you are eroding trust with the business – which will only damage your ability to do your job in the long run. Investing in estimating will help you build the openness and trust you need to be able to get on with delivering to the best of your ability.

Don't develop what isn't needed

As a department, you will have objectives that are given to you as well as those that you set yourself. These may have financial incentives attached to them. If you can deliver certain applications to certain deadlines set by the CEO, you will have done well and may be able to pay a bonus. And if you have some control over how the bonus is distributed, you may also choose to create your own departmental goals and to financially reward your team members according to how well they meet these goals.

Modern economic theory suggests that this is the primary driver for individuals. The reality is not so simple. Money, rewards and achieving goals set by others are all motivating to some degree – but people also have an inbuilt sense of what 'doing a good job' means. Sometimes they will judge this by the technical criteria that they have learnt. More often, they will judge it by how well the software developed meets the purpose of its creation.

Seeing software being used and helping its users work more efficiently, better or with less stress is surprisingly rewarding. One project that I worked on was a web application to replace a spreadsheet system that resulted in an increase in efficiency of 400 per cent. Yes, the numbers sound great, but they weren't as demonstrative of the usefulness of the output as seeing a whole room of call-centre operatives using the system to order their workflow and find the information they needed.

Some software projects don't get this far. They get half-developed, or shelved rather than deployed at the end of development. You might feel that it's inefficient if the business allocates the money for you to develop the software and shelve it, but that the problem is in someone else's department and not yours. This is true!

It soon becomes your problem too, though. Just as seeing your software in production is rewarding and encourages you to continue to do a good job, seeing the results of the hard slog of the past three months thrown away through no fault of your own is soul-destroying. If you let your developers work on projects that are going nowhere, you're the one who is going to have to pick them up and encourage them again after the let-down.

It's not always possible to see the future, but you can maximise your chances of working out what will happen by getting involved at requirements stage. Ask the question. What would cause the software to not be rolled out? What competitor products are there out there that could be adopted instead at the last minute? What else could go wrong? By getting involved right at the start, you might also be able to influence planning and build good relationships so that you know as soon as possible if people's minds are changing.

You might feel that it's not your job to evangelise for adoption. You have led your team to create a piece of software designed according to certain business criteria. If, on someone's whim, the software is shelved, *you* have still surely done all that was asked of you. But if you want to increase your team's sense of value – and perhaps its value as perceived by others – it certainly *is* your job. It might be, for example, that if you can successfully push for implementation of a pilot scheme, misgivings about using the system can be overcome and improvements made once it is in use.

If you're not able get some benefit from the system, your project is likely to feel like a whole lot of wasted time.

What information can be documented?

High employee retention is great for many reasons. One of the most intangible benefits is that you also retain the information that employees have subsumed into their subconscious – the knowledge of how and why to do things, the memories of the changes made to servers, and of what worked and didn't work in the last crisis. For this reason, changing team members can be painful. This is not always immediately visible; instead, and unfortunately, it manifests itself at just the times that you most need the team to be running smoothly.

Documentation is not an equivalent substitute for people who have been doing the job and have built up the knowledge in their heads. On the other hand, you can't assume that the same people will be doing the same job forever. Furthermore, if you've ever worked in a support role or another role that involves almost constant configuration changes, you'll know that, although you can build up a great advantage over someone who hasn't been making the changes themselves, it can still be hard just to remember what you yourself have (or haven't) done.

It can be hard to implement a good documentation strategy, as the benefits accrue in the future, and often to other people. There is no direct incentive to write down what you're doing when the immediate pressure is instead to simply do it. Culture will help – and tactics are needed to kickstart and continue this culture. Make documentation an explicit task, with time set aside for it. Instigate documentation checks. Make a single person responsible for ensuring that documentation is all up to scratch.

A lot of developers will hate this responsibility, but for some it will appeal to their tidy minds. Good documentation is a great way to increase efficiency.

If you're not technical

It can sometimes be easier to earn respect from an individual if you're strong in the same skills as them. If you're a technical manager, you will probably have been promoted because of your strength in exactly those areas that you need your team to excel in. You will also speak the same language as your team.

Managing a technical team without this background, or as a 'customer' from another department, can be an immensely frustrating experience. Some technical practices can seem downright bizarre or contradictory to an outside eye; technical people can be a confident and headstrong bunch (which is only to be expected given how much they're paid…). In this chapter, I'll show you how a non-technical manager can get across this barrier and make sure that everyone is singing from the same hymn-sheet.

- **There is never a substitute for using your brain**
  Process that delivers quality and accuracy in other industries
  can prove counter-productive when the key to productivity is
  ingenuity and creativity.

- **Automate**
  Whenever writing repetitive code gets boring, turn it into
  a more complex piece of code to automate the simple
  repetitive tasks.

- **Who's the expert?**
  How much are you paying these people again? Why are you
  trying to tell them what to do?

- **Train your developers in basic self-management**
  Fed up of small tasks getting forgotten? Don't give up until
  you've tried training basic self-management…

- **Have a plan for working around techies who can't or won't self-manage**
  ... but if it really can't be done, find a way to work around it.

- **Insist on an explanation**
  Technical people react as badly to micromanagement as anyone else – but if you show a genuine interest in understanding the answer, you can always ask the question.

There is never a substitute for using your brain
(Your techies know this – make sure you do!)

In most organisations, when there is a problem with quality or productivity, the answer is to introduce a new process. Too many faults in your widget output? Introduce a quality check or refine the widget-head-putting-on steps. Customers not buying enough? Give your checkout staff an upsell script.

I'm not saying that there isn't a place for process in software development. For any mechanical steps, such as those required to make a software release, process – or better still, automation – is essential. But process can't help with the basic task of software creation.

The reason for this is very simple, and apparent to anyone who has ever done the job (if you're not technical yourself, your team will find it bizarre that you don't understand this). For all that we are working with logical, rational, calculating machines, the process of controlling those machines is a creative endeavour. Trying to do it by rote is like trying to paint a picture by rote. You'll get an outcome, probably much more quickly than you otherwise would, but it quite simply won't be very good.

The processes that do exist tend to be aimed at keeping arduous administration away from the developers, and allowing them to get 'into the zone' – a place where their creativity can be unleashed.

The problems that software developers have to solve are not mundane or routine. There are almost always dozens of different ways of approaching the same problem, and the subtleties of the situation will mean that a solution that may be the correct approach to take in one case could be disastrously wrong in another. So, while you should make sure your developers are aware of best-practice approaches such as dependency injection, you shouldn't be trying to get them to apply them formulaically.

As a developer, some of the least constructive conversations that I've had have been those in which I've been trying to discuss what approach to take with someone who's following a rulebook. One of my colleagues, working on a client site, was terrified of explaining the correct way to do something to the client's tech lead – open debate wasn't welcome, and if he didn't give the answer that fitted with the tech lead's learned architecture patterns he'd be accused of insubordination. The project that they were working on is now three years behind schedule, and still hasn't been released.

Automate

Developers earn their significant salaries because they can bring their intellect to bear on difficult software problems. The flip side of this is that, if the work is repetitive or mundane, developers will get bored and careless or even be tempted to use their untrammeled creativity in unhelpful ways. Luckily there is a solution at hand in almost every scenario. Either there is human input needed because the problem is trickier than it may at first seem, in which case you're home and dry, or the work is repetitive and hence susceptible to being broken down into a simple set of rules. Often, automation of boring jobs is only undertaken if there is a clear-cut cost benefit. But that's only the case if you don't take into account the de-motivating effect on your developers. My rule of thumb is to err on the side of automation, even if there is a slight cost to it. If you need to justify this financially, remember that you never know when you may need to regenerate the same code in another project.

A typical design of many web systems is that each page requires a set of fields to be pulled from the database, copied across the interface, and then outputted into the correct format for the page. To achieve this, an average developer would have to re-implement almost the same piece of code dozens of times, using only a tiny fraction of their brain to do so and possibly even, if sufficiently inexperienced, falling foul of 'cargo cult' programming, copying and pasting code without understanding what it actually meant (see the illustrative example below).

I once saw a zealous and empowered developer taking things into his own hands. Through a highly convoluted use of reflection, he managed to implement some generic classes that meant he had to write the code – albeit quite a lot of it – only once. Not only did the lack of bugs found in warranty deliver a saving to that specific project, but a few months later he found that he had the same scenario all over again on another piece of work. That developer got a good bonus that year.

Illustrative Example: Cargo Cult Programming

During World War II, previously isolated Melanesian Islanders suddenly found vast wealth quite literally falling from the sky. First the Japanese airforce, and then later Allied planes, dropped tonnes of supplies on the islands. The soldiers stationed there received and shared with the islanders goods never seen before, such as manufactured clothing, medicine, canned food, tents and weapons.

After the soldiers left, of course, the goods stopped arriving.

Having no underlying understanding of how these goods had been generated, the islanders imitated the same practices they had seen the soldiers, sailors, and airmen use in an attempt to get cargo to fall by parachute or land in planes or ships. One particularly popular ritual was to perform parade-ground drills with wooden or salvaged rifles.

The goods, of course, never arrived.

In commercial web-based software, a junior developer can often implement a new section of a website by copying and pasting from an existing piece of code. For example, a book-ordering page could be copied from a film-ordering page by keeping fields such as 'Title' and changing fields like 'Running time' to 'Word count'. Copying code in this way is very dangerous; it's susceptible not just to plain old copy-and-paste errors but also to fundamental misuse of code components. To put it another way, how can a developer possibly be architecting and structuring a page to get the best results if they don't even know what commands they're using?

Who's the expert?

You're a busy project manager, CIO or other person tasked with making things happen. You want your technical team to get on and deliver the work that you need so that you can go and report to the board that you've hit your targets. You don't need to know how they do it, or the finer points of why they are having problems now when it was all going well yesterday. Or do you?

I'm sure I don't need to explain why being a cog in a giant machine over which you have no control is demotivating. The opposite of this is having a manager who is interested in what you are doing and why you are doing it. You'll never be able to build up the level of expertise that your team has, but you can use that to your advantage. There is nothing more flattering than being asked to explain what you do by someone who respects how useful it is and acknowledges that they couldn't do it themselves. And of course, building up your knowledge in these areas can only increase your ability to make decisions and communicate them upwards. If you're not from a technical background, taking an interest at more than a superficial level will open up a new world to you. If your senior developer is off, you may be able to help the junior developers by pointing them at the right resources, or sharing a common problem that you've heard discussed. You can have a constructive discussion around issues that arise, rather than an argumentative one. You can do these things, not because you know better, but because you've learnt enough to know how to listen to people who *do* know better.

The other key way to gain respect from people who are your technical superiors is to make it clear how you can be useful to them. Your instinct may be to try and help within their sphere (which they don't need), or to intervene in an authoritarian way (which they might not recognise as help at all). The best way to help is by using your own expertise: management. Can you take away nasty complications with the customer? Can you help to present their skills and recent achievements to help them achieve the promotion that they're after? If you respect and value their space of execution, and show them how it would be useful for them to align it with yours, you'll have an unbreakable bond.

As a developer, I was close to the bottom of the pack. My more enthusiastic, dedicated and naturally gifted colleagues soon outran me. As a result, I started developing along different lines in the support team, where I got more credit for being nice to people than for solving difficult technical challenges. This led nicely into management – and suddenly, a few years later, the techies whose ability to construct elegant, performant functions had dwarfed my own were suddenly reporting to me.

It was a little bit of a shock to everyone concerned. Like most new managers, I was keen and enthusiastic. I didn't stop to think how this might look to people who were already doing the job better than I could.

After a rocky start, I soon realised that only by making the effort to make myself useful would I come to be regarded as such. 'Helping' people to rationalise their risk logs and follow new processes didn't seem to them like help at all. So instead I started to focus on how I could actually be helpful. I took away troublesome admin tasks. I helped open up new opportunities that the developers were trying to move into. I asked the question 'How can I be of help?' And I didn't stop until I got an answer.

Train your developers in basic self-management

Technical people are not known for their self-management skills. They are clever enough to write intricate code that can control spaceships, but they can't remember to send an email on time. Technical people are good at concentrating hard for hours, getting lost in their work, forgetting to go home – and pulling it all out of the bag at the last minute. They are not good at planning ahead, or stopping to replan rather than working their way out of a crisis.

Accepting this apparent fact can seem like a sensible conclusion. It's easy to think that your technical experts aren't capable of self-managing – that their brains don't work the right way. Sometimes you may need to accept this and deal with it. Most of the time, however, you *can* help developers to improve their self-management skills, if you explain clearly what they need to do.

Self-management most often fails when you don't have a suitable system in place. The first thing that I do if someone is not delivering what we agreed, when we agreed (because they 'forgot'), is to examine what their process is for remembering to do things. Sometimes they don't even have a process. They are trying to remember it in their head, or they wrote it down on a piece of paper – which has joined a lot of other pieces of paper on their desk.

In order to manage more than one task, you need a single, consolidated 'to do' list. You need to know where to look when you need to find out what to do. What's more, your list needs to contain *all* the information that you need – and you need to check it regularly, to see whether you need to be planning now for something that you need to achieve by Friday.

A central, soft-copy 'to do' list is a well-known self-management tool. Most people who have problems with getting things done already have a 'to do' list. (Of course, if they don't, implementing one is the first step to self-management!) Clearly, as with a risk log, having a 'to do' list isn't sufficient on its own; you also need to use it well. I've come across three main areas where 'to do' list usage breaks down.

1. Things don't make it on to the list

If you have a list, and you use it well, it will work almost automatically. But there's one step that's open to human error, right at the start: if something doesn't make it onto your list, it won't get done unless you happen to remember that you didn't add it.

As technology advances, we can get nearer and nearer to the point where you can carry an electronic 'to do' list around with you and add things as soon as you discover you need to do them (in fact, by using a cross-platform note application like Evernote, we are already there: voice technology and better interfaces will make these tools more and more crucial to our jobs). But in the meantime, you need to make sure you write things down in a mini-list that can be transferred into the main one – and always prioritise putting things on your 'to do' list over any other task!

Lots of task systems – the task list that comes as part of Microsoft Exchange, for example – employ the concept of a dated task. A date is, of course, an essential part of a task, as it tells you when you need to get something done by; if you have a task without a date, you can end up deferring it forever.

Most people use this date to set the deadline by which a task needs to be completed. This is the most natural date to use; it's the important one that you don't want to forget. You can set a reminder to tell you about the task – say, three days in advance of the deadline – giving you plenty of time to deal with it. But most of us don't have the luxury of just being able to drop everything and start on a new task when we get a popup from our 'to do' list. So you need to dismiss the popup, or tell it to return again later. Soon you find yourself managing multiple popups at the same time. This is the opposite of what you set out to achieve by having a task list. The idea behind the list was that you would have one consolidated place to go and see what you needed to do next – now you have a list, and several popups!

The second most common reason why people drop actions (after not having a list at all) is setting a reminder date that's so late that, by the time it pops up, it's too late to do anything about the task they need to get done! I find that I need up to three dates on any one task – the agreed deadline, the date that I plan to do the task, and the latest date by which I can start the task and, given my other commitments, still have it done before the deadline.

3. Emotions get in the way
Managing a task list feels like the driest activity that one can undertake. Surely it's not one where emotions play a part at all? To my surprise, I discovered that it is.

I have a habit of allowing low-priority actions to build up if I can't fit them into my day-to-day work. These can range from writing a new managers' newsletter to improving some documentation or policy. One year, I worked the three days between Christmas and New Year, and with everything else powered down got through my entire low-priority list without breaking a sweat. This means that the tasks that had been sitting looking at me accusingly over a whole year only took three days to completely resolve. That's an average of less than five minutes extra every day! Why hadn't I been able to do them sooner?

On closer examination, I found that I had got through plenty of *other* low-priority tasks – namely, the ones that I found exciting and that I was enthusiastic about. It took me a long time to realise that the tasks that I wasn't getting done were ones where something was blocking me from doing them during my work week. They were tasks that I really didn't want to do, or that I wanted to achieve but didn't relish working through, or where I didn't know what I needed to do next (so it was easier to defer them than to sit down for five minutes and work it out).

It's important to think carefully about the things you have trouble actioning. You may be able to work out the root cause, or you may not – human beings are funny creatures. But what's important is that you develop ways of 'tricking' yourself into doing what you need to do, and that you keep trying different tactics until you find some that work.

In one productive chat that I had with a developer who was facing task issues, we took just ten minutes to solve a problem that had been plaguing him for years. He preferred development to admin tasks (who doesn't!), but he knew that he had to get through the admin and so added everything to a task list. When the admin tasks got to the top, he would do them. That was the theory, anyway

– 'But Zoe,' he said, 'even when the admin tasks get to the top of my list, I just do development tasks that are lower down.'

We discussed the reasoning behind his list and tried to understand what caused him to do certain tasks. Essentially, he considered everything on his 'to do' list as something that had to be done (although the top tasks were more urgent). This meant that the development tasks on his list were, according to his rules, 'fair game' – even though the overarching aim of his list was to get him to do the smaller, less fun tasks that he would otherwise forget.

This made the solution simple. In order for the developer to get his admin tasks done, he had to take everything else off his list. At that point, he had no choice but to do them.

This may sound extreme (and somewhat contrary to the point of having a list), but in fact I find it a key process. Once you start training yourself (by whatever means) to do the things that you have an emotional resistance to, it gets easier. As with everything in life, if you want to get better at something, practise it. I once read a book in which the author suggested that, in order to start doing the things that you don't want to do, you should do two such things each day, for practice. I have 'Do two things I don't want to do each day for practice' written at the top of my 'to do' list. I'm not only now much better at getting on with the things that achieve a long-term aim but aren't things I want to do in the short term; I can also enjoy tasks like tidying up the kitchen by thinking of them as practice at doing things that I don't want to do.

There is another approach to managing techies who aren't good at self-management, which is to take away the problem instead of training them to get better. Some techies would choose this option, and will criticise you for forcing them to do extraneous tasks that don't play to their core skillset. I think that this is the easy way out – and that's rarely the right option. I believe that being able to manage yourself to achieve what you want is one of the most important things that you can learn, and it's something I think everyone should have the opportunity to learn.

Have a plan for working around techies who can't/won't self-manage

I don't recommend it, but it is possible to be a developer without learning how to manage your own tasks. If you have a project manager or project assistant who is willing to keep track of what needs to be done and to feed tasks into you when needed, the whole team can still work.

Of course, as per p59, I think you should try to train everyone in self-management in the first instance. It's possible to be a very good developer without being able to manage, but it's not possible to progress far down any career path – e.g. towards tech lead – without learning this.

If you've tried and failed to build a team of developers who can all manage their own tasks, the solution is to split the workload by having more support staff. If you need to manage around developers, you will probably need to hire extra project administrators to keep track of what tasks need doing when, and to manage all administration. This will give you some extra developer time, but the overall cost could well be higher.

Insist on an explanation

In *Respect your team* (p16), I talked about why you need to give clear explanations to your team in order to increase buy-in and, of course, to show that you respect your team enough to let them in on your decision-making. One great advantage of this is that, if you are in the habit of giving explanations, you are also in a position to demand them when you are unsure about exactly why something has to be a certain way, or why it should take so long.

As a manager, who probably doesn't often have the time to go into the detail, you need to be able to assess proposals and plans at a high level. Some transactions can just be done on trust: with your longest serving managers, you can get to a situation where they have such a history of delivering on time that you can let them put together plans and simply look at whether the cost works for you or not. Even then, there may be times where you need to clarify exact approaches or establish why a certain approach doesn't work.

You should develop a healthy suspicion of being given the 'right' answer. If you get the answer that you expected (or were hoping for), it might be that everything is tripping along swimmingly – or it might be that this is the obvious answer to stop you enquiring further. It could also be that, in your rush to assume that there are no crises for you, you don't hear the substance of what is being said and interpret the answer according to your own expectations. It's too easy to hope for and assume the best. Look for further details and reasons to ensure that you're not being misled (accidentally or deliberately).

If you're uncertain of something, blind trust is not the answer. Good, challenging questions and a culture of explaining what you're doing and why will help you to get to the best answers.

It's all about people

It's fair to say that many people choose to work with machines precisely because they prefer not to deal with people. But once you hit a senior role, it's no overstatement to say that your role is *all about* the people. If you'd rather it wasn't, your job will be an uphill struggle.

You will encounter people when trying to deliver on every facet of your job description. Meeting client requirements will mean you need to sit down with clients; hitting delivery dates will entail working with third parties such as hosting providers; to deliver anything at all, it will be essential to have great relationships with your team.

For this reason, this is one of the most important chapters in the book.

The first two sections will deal with the customer, the third with people outside your organization, and the rest with your team. Basic people skills apply to everyone you come into contact with, but some more specialised skills can be more useful, depending on who your target audience is.

- **Be nice to the customer**
  If 'be nice to the customer' doesn't sound like an essential commandment to you, you *definitely* need to read this section.

- **Make friends with the enemy**
  Why not make sure that you and the team who will use the software are on the same side, rather than fighting against each other?

- **Networking people, rather than PCs**
  An oft-maligned skill, networking is the one area in which it is absolutely essential that you are on top of your game.

- **Internal networking**
  There's more to networking than targeting powerful individuals who could become clients or employers; internal and downward networking is just as important, if not more so.

- **Make issues into challenges and puzzles, rather than threats**
  Everything difficult in life can be viewed in two ways – we embrace challenges, and run from threats.

- **Believe in everyone**
  If you don't believe in them, how can they believe in themselves?

- **Genuinely care about people**
  It is possible to genuinely like everyone, if you take the time and effort to find some common ground – and it's essential if you want to be able to relate to them.

- **How to make people happy**
  In the 21st century office, this is every manager's key job responsibility!

Be nice to the customer

Directing your efforts to help the customer won't be a new concept for you. This is the case whether you are a software services company, where the customer is another organisation, or an internal IT team where your customer is the rest of your own organisation. As an output-focussed senior manager and leader, you know that a good relationship with the customer is absolutely key to delivering good service. Further, as it is the customer who is judging your outputs, you actually need them to be involved in setting the goals that define what success looks like.

So you will be surprised and horrified when your techies seem to get into a combative relationship with most, if not all, of the external decision-makers. If the customer agrees with every suggestion or is happy to be uncritical of technical decisions, there will be peace. But if the customer is working to constraints that they don't communicate well, and pass these on as ultimatums, the technical team is unlikely to respect this approach and enter into discussion to get the best solution; instead, they will try to work out how to impose the plan that they already intended to implement.

''You want one server; OK, on your head be it,' one consultant wrote on Twitter recently. 'I shall be adding a disclaimer in the design document to say all decisions made outside of my solution is no fault of my own etc. if it breaks.' It's an all-too-common cry.

The problem is that your techies are guardians of expertise that the customers don't have. Because their focus isn't on communication, they assume that if the customer wants something different they must just be stupid or pig-headed. If you have business analysts, requirements analysts or product owners, this is a good time to get them involved in bridging the gap. Failing that, you may need to step in yourself and take over. But if you haven't trained your technical staff to help out the customer (i.e. to think: 'They're so inexperienced they think putting it on one server is a good idea. They need our *help* to not make the wrong decision'), don't assume that that they'll understand good customer relations in the same way that you do.

Make friends with the enemy

It seems that there is no technical department in the world that co-exists peacefully with the rest of the business. Technical people think that the business users are being uncooperative when they're unable to grasp the simple principles that explain why things must be done a certain way. And they are always changing their minds! Business people find the technical team sullen and uncommunicative, and often, just when they think they've defined in minute detail what they require, at countless long and boring meetings, they find that a simple feature that was obviously required for the system to serve any purpose at all has been omitted and that they need to wait weeks for it to be re-implemented.

Alliances will form naturally along these lines, with business analysts caught in the middle and fighting against both sides. This is a perfectly natural outcome, given the pressures on and expertise of both parties – but it doesn't have to be this way.

To stop this situation arising, you need to have explicit tactics. There are three ways to get the business team and the technical team working together effectively.

1. Cross-socialise

Team-building events, whether horrific raft-creation exercises or just outings to the pub, are common practice in most workplaces. Even when events are not explicitly organised, socialising occurs naturally – after a hard week, everyone will tend to head out together in order to relax. This kind of informal socialising often builds teams much more effectively than pre-planned exercises.

People naturally socialise with the people with whom they work most closely (or even those to whom they sit nearest). But if you can engineer situations where your team and the team that they are building software for socialise together, they will be better able to understand each other and work together towards the end goal. It needn't be anything formal – you could simply suggest to the usual social outing organiser (there always is one) that they invite the product owner and end-users the next time they go out. You'll also need to discourage technical topics of conversation – otherwise you'll just have two groups socialising separately in the same venue!

2. Find common objectives

At the end of the day, everyone has the same objective: to deliver an excellent piece of software that serves the business's needs. So why does it always feel like a battle?

Focussing discussion on the things that people have in common, rather than those that separate them, is a good way to encourage the wider team to pull in the same direction. Establish general principles that both sides believe in, such as low total cost of ownership, excellent user interaction and differentiation from competitors. When disagreements arise, refer back to the agreed principles to help guide everyone towards agreement. Discuss how spending longer on code structure will lower the total cost of ownership; don't just assume that it's obvious to everyone. Explain how the system will be used, and talk around the options rather than assuming that the developers will work it out for themselves.

3. Try to see things from the other point of view

One of my favourite sayings is 'Walk a mile in another man's shoes – then you'll be a mile away, and you'll have his shoes'. Trying to imagine how you would behave in another's position can bring several benefits. First, you may realise *why* they are pushing for a particular solution. Do they have pressures imposed on them by others? Are they going to be judged negatively if they can't achieve a certain outcome? Second, you may be able to think of a better solution to their problems, which changes the position from one of adversity to one of persuasion – they may not agree outright that your solution better meets their objectives until you link it back to the things that they care about. Finally, appreciating both sides of an argument might lead to new insight, and reveal a 'third way' that, instead of forcing an unsatisfactory compromise, enables both sides to achieve a productive outcome.

Networking people, rather than PCs

As a CTO or department head, your role is not a technical one – even though you work in a technical area. You're responsible for relationship building, both internally and externally. You'll be used to attending conferences in the fields in which you work, and building networks of people with whom you will work in the future, as suppliers, customers or partners.

You'll probably be in and out of the office on a daily basis. Your developers will remain almost constantly at their desks, whether they work in the office or at home. This is absolutely as it should be – you can't write code when you're out and about.

But developers could do with learning just some of the skills that you use every day. Many of them will attend conferences or workshops in order to keep up-to-date with new technology – such events also provide good opportunities for them to build links with other organisations, which will benefit both them and the department.

If you ask around the technical peeps on your team, you'll find two different views of networking. Many developers – those who don't even talk to the sales team – will not really be aware of networking as an activity. They'll know that some people talk more to others at events, but will assume that either they know more people or are more naturally chatty. It therefore follows that salespeople will do more of this as they are by nature 'people people'. Those developers who *have* heard of networking, meanwhile, will consider it to be a sales activity – and will probably be terrified of it.

Both of these viewpoints lead to the same conclusion: that networking is an activity for salespeople, not for developers. What's more, most developers will believe that this is something that they are inherently not very good at. This is not true! As someone who has made the crossover between technical work and sales, I know that networking skills *can* be learnt. It's not easy, and it takes hard work and determination, but it can be done.

For anyone looking to polish their networking skills, I thoroughly recommend the book *Never Eat Alone* by Keith Ferrazzi. As long as you can manage to not feel inferior to Keith, who manages to work 20-hour days and never misses a chance to make a new connection, the book is a great source of ideas for improving your networking ability. Keith himself, the super-connector, insists that he is not a natural networker, and that his skills in this field had to be learnt. There's hope for all of us.

There is also my own second book, "Networking Know-How" in which I expand on my thoughts on networking in more detail.

I've found that a common answer to the question 'What is the point of networking?' is 'To make sales'. This answer ignores the wealth of opportunities that are available to developers via networking. To help you convince your developers to give it a go, here are just a few of the potential benefits.

1. It can be hard to know which are the best talks to attend at a big conference. Other delegates might have seen some speakers before, or have some useful ideas about who is worth seeing.

2. Obviously, you're attending the conference in order to learn about new technology or new ways to use it from the speakers. But you'll be able to learn about *other* ways to find these things out – maybe through recommendations of books to read, forums to join, or even other conferences to attend – from other delegates who are in the same position as you.

3. You may even be able to learn directly from other people at the event. Remember that the speakers themselves are often around before and after their talks – what better opportunity to ask

a question that's been puzzling you? Other luminaries of the technological world may also be there – you won't know unless you get chatting!

4. A conference is one of the best environments to meet new people who share your interests. If you're working on a personal project or website, this may be the perfect place to find collaborators who are interested in the same things as you and might want to get involved with your projects. Conversely, you might meet people with exciting projects of their own that you can get involved with.

5. Even in the world of developers, there are experts and trailblazers. This may appeal to some developers more than others, but if you're getting your name mentioned at events as someone who knows what they're talking about, it could be helpful for your career – whether you end up looking to launch a book on a particular topic, or just looking for a new role.

6. Finally, if you talk to other people rather than standing in the corner feeling nervous, you will quite simply have more fun while you are at the event. You get to hang out with interesting people, and talk about your favourite topics! I think this is the key benefit that I now enjoy as a result of having worked hard at networking.

So those are the 'whys'. How about the 'how'? How exactly do you go about taking the plunge and talking to strangers – especially if you're used to mostly interacting with a machine all day? These are my top tips for developers who are looking to try to improve their networking skills.

1. Jump in

In networking, as with many other things in life, the hardest part can be just jumping in and kicking it off. That's why you need to practise! Set yourself as high a target as you can tolerate (but one that's still achievable) for the next event that you go to, and determine to talk to that many people – five or ten is probably a good starting number. Then, each time you're at a loose end, look around and pick someone; ready yourself with a pre-prepared icebreaker question, and set off. When you're done, tick one off on your list. See, that wasn't so hard! Increase your targets for future events until you get into the swing of it. Then you'll find you end up instinctively talking to people before you even realise you're doing it.

2. Find the popular people

If you're nervously looking for people to talk to, it can be really easy to skirt around the side of the room and look for people who are standing on their own and look as nervous as you feel. This is surely more likely to get you someone to chat to than charging into the big group already in conversation, right? Wrong. If you talk to the people who look like easy pickings, you're actually making life harder for yourself. The easiest people to talk to are the people who are better networkers than you are – and you can be sure that they will already be in a conversation. Don't be put off by the fact that groups of people talking look like they all came to the conference together, met by pre-arrangement or have been talking together all through the event. Most people will be in exactly the same boat as you. They will have just met, and will be keen to meet more people. If you can subtly join the edge of a big group, with an opening line that either simply acknowledges what you're doing ('Hi, do you mind if I jump in here?'), makes a non-committal comment or responds to an existing conversation, you'll be part of the group in no time. And once you're part of a group of more than one person, you have ready-made introductions to several more people (who you can check off your list!).

3. Line up some ice-breakers

It is very hard to immerse yourself in something if you don't know what you're doing. So go prepared. Have some questions lined up, memorised, that you can ask anyone at the conference. Simple ones include 'I'm Zoe, who are you?', 'What company do you work for?' and 'What's your role?', but you can make it a bit more interesting by finding a more non-standard opener: 'What's your favourite coding language?', 'What do you think is the best benefit of Agile programming?' or 'What app would you most like to have on your phone that hasn't been written yet?' might all make effective ice-breakers.

4. Know what you do

If you worked in sales, or in a job that required extensive networking, you'd have a short answer prepared to the question 'What do you do?' This is known as your '118' or 'elevator pitch' – 110 seconds being the average amount of time it takes to ride in a lift between floors, and 8 seconds being the average attention span of a human being. The concept of an elevator pitch is to imagine that you have got into a lift with someone who you would like to impress, and you have the time it takes to reach their floor to impress them enough that they want to meet with you further. This concept still applies if you are a developer! What's interesting about your role? What might people want to hear about? 'I design and code applications to help deliver the world's news to the people' is much more interesting than 'I'm just a developer at CNN, I guess'.

5. Ask people about themselves

My favourite business-book author, Dale Carnegie, provides the best answer to how to deal with this situation in his seminal work *How To Win Friends And Influence People*. Dale suggests that you always ask people about themselves. This is not just the topic that they are most knowledgeable about, but also the one that they most like discussing. Avoid telling stories about yourself or trying to find topics that are of interest to *you*; instead, try to understand more about *them*. What do they like about their job? What interests do they have outside work? What's the most exciting project that they've ever worked on?

An extension of asking people about themselves that's very popular in modern networking best-practice is to look for ways in which you can help every single person that you meet. Do you have a contact that might be useful to them? Do you know a book that they might find helpful? This can be a good technique for taking the pressure off if you're worried that you are interrupting or disturbing people. If you can make it work, you're well on your way to being an excellent networker.

6. Know what questions you want to ask

If you're going to reap all of the benefits that I listed earlier in this section, you'll need to know how to get the information you want. Decide in advance of the event what you would most like to gain from it, work out which attendees are most likely to be able to help you, and have a list of questions at the ready.

7. Get their details

If you make a good contact at an event, the chances are that it will be useful for you to talk to them again. Make sure you carry business cards. If you don't attend many events, moo.com is an excellent site for printing small quantities of cards at a professional quality. Moo can be used even in large organisations, for example to top-up single sets of cards when someone joins the company, or to order cards at a quick turnaround time and low minimum order quantities.

Remember that, although you carry business cards in order to give them out at the event, the most important thing is to make sure that you get the card of the person with whom you're speaking. Nine out of ten people won't get in touch after exchanging details (even if they promise that they will!), so you need to make sure that you have their details so that you can call or email them. A new tactic that I use with people who don't have cards is to fire up my smartphone and connect to them immediately on LinkedIn. It's the business card of the future!

One thing that trips up new networkers is that they spend ages working up the courage to go and talk to someone, only to find that they then don't have the courage to *leave* the conversation! You may well find that you get into a conversation that shows no sign of ending. It may be extremely interesting or it may not, but your target count of people to chat to is there to remind you that the aim is not to have a good chat but to meet different people. If you feel that you have just pushed someone into talking to you, it can feel dismissive, even rude, to then make your excuses and move on – but it's essential that you learn how to do this! Again, it helps to have some pre-prepared lines ready. The best can be the most direct: 'I need to make sure I meet some more people before the end of the break' or 'I'm off to work the room'. Or how about 'I'm actually working on my networking skills and the book I just read suggested I set a target number of people to talk to – I'm on two and I need to get to ten, so I need to meet some more people! Let's catch up at some point after the event'? And if you're not ready to bare your soul, I find 'human need' excuses work well – that you need to use the bathroom, or get another cup of coffee.

Online networking is a different skill, but it's one worth considering for your developers. LinkedIn is commonly known as a tool for re-cruiters and salespeople, but a lot of technical 'groups' have sprung up, which in effect create chat forums on technical topics from .NET to Ruby. With the recent addition of the skills section to LinkedIn, you could try searching for people in your network with highly-rated skills in the technical areas that you're looking for help with.

Twitter, meanwhile, is also well-known as a great tool for getting tech updates and, many well respected technical superstars, from Jeff Atwood to Joel Spolsky, share their knowledge on the network. I've yet to find a genuine business use for Facebook, but wherever there are people, there is potential!

Internal Networking

It's always been recognised that the people who get ahead seem to be the ones who know 'the right people' within an organisation. While this is sometimes regarded as a skill, it's not historically been considered a skill that can be learnt. More often, it's been dismissed as unhelpful and unfair nepotism. As business theory has developed, learning how to build your network both externally *and* internally has come to be seen as an essential skill – and for good reason. In our ever-more-interconnected world, *who* you know is now much more important that *what* you know.

To this end, the term 'internal networking' has become widely used within large organisations, particularly among ambitious young professionals, and I'm convinced that if you work on your skills in this area you'll soon see tangible business benefits. Sometimes, even people who recognise the need for – and hence do well at – external networking don't consider internal networking to be at all important.

Where internal networking *is* practised, it's often seen as a rush to see who can take the CEO out for lunch. In reality, networking downwards rather than upwards is likely to reap just as many rewards (if not more). This is because the success you achieve is entirely dependent on how well you attract and retain the best staff.

Networking is in essence about making people aware of what you do – for example, what products and services you offer, what common interests you share and can help each other with, what function your team performs, what roles you're recruiting for and what requirements you need help with.

Most people start to think about networking when they have an immediate requirement. But networking is much more beneficial if it's treated as a longer-term activity – if you put out a job advert internally when you already have a strong network, you'll receive a much better response than if you put it out 'cold'. You also run the risk, if you network for immediate need, of being seen as someone who is always asking for something. It's much easier to make a genuine contact for the future if you're coming to the relationship with no ulterior motive.

There could be all kinds of reasons why you might want people to know about you internally. I guarantee that at least a few of the following will apply to you.

- You want to recruit internally – in which case you'll want to attract the best people in the organisation. Find who these people are in advance!

- You need to draw on expertise from another department, or to borrow resource temporarily.

- You're waiting on something from another department that isn't being delivered, and you need a contact to call on to find out the real reason why.

- Another department is about to do something that would invalidate the work you have done recently.

- You're in a position to help another department.

- You're able to find out information that could be helpful – why your manager is asking for a specific deadline, say.

- When people discuss your department and your work, you want them to be talking to someone who knows what you are doing and can give a positive answer.

- You've found out about other opportunities in the organisation that suit you or someone that you know.

- It's fun to meet new people!

As a developer, although you will certainly benefit from applying the networking techniques explained in this book, you may be able to get away without doing so if you're prepared to remain a coder for your entire career. Once you reach a management position, however, your entire role could be determined by your network. Make sure it's a good one!

Make issues into challenges and puzzles, rather than threats

In technical work, problems happen. If they didn't, you could replace your crack development team with the automatons that everyone else suspects them to be. It's the problems that mean you need clever, inventive people around to fix them.

When you're working as a developer, this doesn't always seem so clear. You have a piece of software to build and a plan for how to build it. You can easily feel that unexpected obstacles are going to derail you.

But your best people thrive on problems. Some want to be seen as a 'fixer'; they want the respect that comes from having others know that they can step in and sort everything out. Even better are the peeps who love the problem for the sake of the problem. If the development environment isn't playing nicely with the new library you downloaded, the best response is 'That's interesting!' (Rather than 'Not again!' or 'I wanted to get this software shipped tonight!').

Self-belief is an important part of this. Your peeps need to believe that *they* can solve the problem. Not their pal, or someone on the other team, or someone at the other end of a forum. Themselves.

A personal breakthrough for me when I was working as a developer was moving to Australia. I still had support overnight, via email or in calls at the end or start of the day, but if I wanted to get something done *now*, I needed to sort it out myself. The first time I hit a problem, I tried a well-known technique. I went to make a cup of tea. It had always worked for me in the past: make a cup of tea, have a cigarette, stare blankly at the code… and soon I've spent nearly two hours struggling with the problem and feel perfectly justified in asking for help. But it didn't work in Australia. I made my cup of tea, came back to my machine – and the problem was still there. I didn't have any choice but to sit down and start thinking about it.

A funny thing happened once I'd got to the end of my cup of tea. I wasn't thinking about tea any more, or staring idly out of the window. I'd been gripped by the problem that I needed to solve. I was thinking of answers and verifying solutions. It took me most of the day to solve an issue that one of my colleagues could have sorted in 10 minutes, but I learnt the most important lesson of my career. I could do it myself.

So watch out for the developers who let other people solve their problems. Work on plans with them so that they can build up the confidence to do it themselves. Take away the easy route – they'll thank you when they've made it as a tech lead two years later.

Believe in everyone

One of the most helpful pieces of advice in *How To Win Friends And Influence People*, which is still very much relevant after 80 years, is to make sure that you like people. On the surface this sounds like crazy advice. You can't like everyone. Some people are just different. Some people are downright unpleasant, and it's not your job to work around that.

The thing is, as a manager, it *is* your job to work around that. Obviously, if you're taking your recruitment duties seriously, you would hope to recruit people who you respect and trust and want to work with. But sometimes you'll need to work with someone from another team; sometimes a 'bad egg' will slip through the recruitment net. And unless you work for a phenomenally successful software house, you have limited control over the people you build software for.

Is it possible to simply learn to like everybody? After challenging myself to do exactly that, I can confidently say that yes, it is. But you may need to rethink your basic understanding of people to get to this point.

The first concept you need to get to grips with, which I always knew but never quite fully comprehended, is that people who you think are unpleasant, or even downright bad, don't think that of themselves. To give you the kind of example that really brought this home to me: Joseph Stalin didn't think he was an evil mass-murderer. But I won't go too much further with this analogy, as hopefully you'll never have to learn to like a perpetrator of mass war crimes. An example that's more likely to be relevant is that the arrogant, know-it-all techie, who seems to exist only to irritate others, probably doesn't see herself that way at all. The fact is, if you need to get on with someone and you have an inflexibly negative opinion of them, you will find it hard to get their buy-in to the decisions that you need to cascade through the organisation.

Step one towards liking someone is to getting to know them. Find out what has affected them in the past, what they want and what motivates them. I have found throughout my life that people who behave badly are invariably unhappy. Happy people are able to have the generosity of spirit to be nice to others.

Sympathy is key to liking someone. The principle is clearly expressed in the episode of *The Simpsons* where Marge Simpson has to paint a picture of Mr. Burns, the evil owner of Springfield's nuclear power plant. Marge finds she cannot paint Mr. Burns without finding something to like about him. In the end, she compromises by drawing him as a frail old man – she finds that even if she cannot like him, she can have sympathy for him. Mr. Burns, being a cartoon character, can approximate the concept of irredeemable evil that we all hold. In real life, of course, people are not like that. Everyone whom you dislike for some reason will have other characteristics that you admire. You just have to find them. You'll naturally socialise with people whose company you enjoy – but try making an effort to talk to people you wouldn't usually sit with. You are likely to find something in common, even if it's a technical topic.

In my attempts to get to like people with whom I work, getting to know and understand them has been the only step necessary. If you find this hard to believe, the only thing to do is to try it yourself. And, of course, trying new things is another end in itself.

If you're responsible for someone's career development (and if you run a department you should consider yourself responsible for the career development of everyone working there), you need to go one step further than liking them. You really need to believe in them.

It's not true that everyone can do everything *right now*. But *right now* is an important caveat, because the more you think about things you can't do as things you can't do *now* but could do in the future, the more you'll see paths that can get you there. I can't parachute jump, but it would take a short course to do a single jump and a longer training program to become a proficient amateur. If I set my sights higher, I could enlist in an organisation that requires people to parachute jump for a living. If I look at a scale of ten years, I could be a professional.

At the same time, you must avoid the temptation to give people false hope. If someone struggles with management tasks, don't promise them that they will be a project manager by this time next year.

The only thing worse than someone who doesn't believe in you is someone who pretends to believe in you and then reveals at the last moment that they didn't after all.

Instead, think clearly about the first steps that need to be completed before you can get to the end goal. Discuss these as initial goals. Try to explain that everything in life takes time, and that major achievements are built on top of smaller building blocks. If you're talking to someone ambitious, you need to try and nudge them away from the idea that everything is a competition and that they will need to *get there first*. The only helpful way to proceed in your career is to set your own realistic goals and work to achieve them. Being better than others is not a career goal.

Work out a plan that you confidently believe your managee can hit. It may be that you are only confident that they can hit it if they achieve certain sub-goals. You can be open about this. Most underachievers suffer from a lack of belief in themselves. If you can provide some of this belief, it will help them on the way to where they want to be.

Genuinely care about people

We have a large amount of written training material to help people understand how to be a good people manager. I recently condensed this into just two maxims: 'Genuinely care about people and want to help them', and 'Be in a position to be able to do this'. The former speaks for itself; the latter is an exhortation to train yourself to be as helpful as you can, and to challenge yourself to find out how to solve difficult situations. If something goes wrong, analyse it and learn; if you feel out of your depth in a situation, seek help and learn so that you can help your managee. To my mind, this is just an extension of the usual best-practice continual improvement self-training that applies across everything you will ever do.

I think that genuinely caring about your managees is therefore the key thing to do. If you want success and development as much for them as you do for yourself you will go about your job in an entirely different way. Imagine if you could go back now to when you were starting out in your career, but keeping all the knowledge that you have built up over the years. How quickly could you progress? What new heights could you achieve? As a people manager you have the opportunity to achieve something similar by making your knowledge and expertise available to those more junior, who can use it to deliver more for the company.

The complication is that they probably won't want exactly what you wanted, and may not agree with you on the best way to go about it. If you are genuinely trying to help, rather than trying to persuade them to take on a difficult project that no-one wants or to fob them off when they say they want to progress, it will be much easier to help them. If you're actively finding options for them to try and roles for them to aim for, and being the person who believes that they can do it, your managees will develop more quickly.

As a leader in the department, it's your job to make sure that this is happening for everyone. If you don't make sure everyone has the best support in trying to make the most of themselves, you'll find that your department will stagnate. Staff turnover will increase as your best developers leave, and quality will go down. There is no single better lesson that you can learn.

How to make people happy

Hands up if you consider it your job to ensure that everyone under your stewardship is happy.

If I asked this question to a room of CIOs, I would expect to see a sea of hands. With the advent of modern management techniques, we understand that happy and engaged employees do better work and stick around for longer. Of course, it's not always easy to make this happen. Sometimes, ensuring that every single member of your team is happy can appear to be in conflict with your other responsibilities, such as delivering a great service and making the bottom line add up. But it doesn't have to be.

It's important to consider what we mean by 'happiness' in this context. In popular culture, happiness is now often equated with hedonism and other forms of instant gratification. This is not the type of happiness that it's your job to provide, nor is it any use to you in achieving your objectives. You're looking to engender the following types of happiness.

- A feeling of belonging

- A sense of being respected

- Challenging and fulfilling work

- A career plan that means that the work you are doing now will lead to personal development and the ability to meet your goals in the future

This is not an exhaustive list, but it illustrates the kind of happiness that I'm talking about. Thinking of happiness in this way, you can see that the real value of freely available chocolate biscuits and access to fussball and other amenities is not in the short-term pleasure that people get from engaging with them; they are valuable because they are evidence that employees' happiness is valued, and that they are respected enough to be allowed to manage their own time.

So how to achieve this ultimate state of happiness? The high-level answer is simple: you need to find out what people want, and you need to provide it. The *useful* answer is a lot more complicated. The distinction between short-term gratification and longer-term happiness is not always apparent to everyone. The classic example – which will be recognised by anyone who has tried to lose weight or just to keep healthy – is the difficulty in deciding between the food you want to eat now and the bodyshape and lifestyle you want later on.

Simply asking people what would make them happy, or happier, is likely to elicit short-term happiness responses. Instead you need to engage with what they really want to get out of life, and find out how much of that you can help them to achieve within your organisation. I won't pretend that you won't sometimes end up shooting yourself in the foot this way: helping people to achieve their dreams may mean that you help them to be somewhere other than your organisation. I think that this is an acceptable trade-off in return for committed, engaged and happy staff, and for a great incentive to retention for everyone else.

When I've stayed with an organisation for a long time, it has been precisely because I have continued to have the opportunity to learn. When I've been learning skills that make me very valuable in the marketplace, this has increased my ability to jump ship and work somewhere else. But a learning culture and great environment removes incentives to leave as long as I'm continuing to learn – and enjoying myself while I do so!

The flip side to identifying what makes people happy enough to stay is doing your best to find out what makes them unhappy enough to leave. Implement an exit interview process, to give leavers the opportunity to tell you freely, without fear of repercussion, exactly why they don't want to work with you any longer. Try to set up the exit interview with a person who isn't likely to be personally insulted by the feedback. Usually the employee's line manager will be suitable, but, if there have been disagreements with line management decisions in the past, you might want to get someone else to conduct the interview. Remember that the exit interview is for *your* benefit, not that of the person leaving. They may not want to invest their time in giving you tips to improve your organisation, so make it easy for them to do so. Encourage the hidden desire within everyone to point out things that are wrong or to get unpleasant experiences off their chest. Showing a desire to act on the information given, although not directly of any use to someone who is leaving, will reinforce the idea that they are doing something worthwhile. Whatever you do, you absolutely must not allow there to be any negative consequences of people being completely frank with you. Don't get upset, don't contradict them, and don't let yourself find an excuse to cut their bonus if they tell you something you don't want to hear.

And finally, don't wait until someone leaves to get this information. Encouraging people to air their grievances without repercussion while they're still with your organisation may help you turn some unhappy people into happy people who see their future as being with you rather than somewhere else.

Management

Management is the nuts-and-bolts job of making sure that everything is working on a day-to-day basis. For some people, management is all the job entails, and the critical functions of leadership and vision are overlooked; for others, management is dismissed as routine and boring, and sacrificed in favour of more glamorous and exciting activities such as strategising and expansion.

The truth is, as always, in the rather prosaic middle ground. Management and leadership are both important. Management without leadership means that an organisation can deliver brilliantly without knowing what it is delivering; leadership without management results in an incoherent scramble where, despite a lot of great plans and ideas, nothing is delivered at all.

Management is a skill that at its best is intuitive and light-touch. Great managers plan ahead, for sure, but they can take the decision to diverge from the plan, and seem to magically know what problems are going to materialise before they actually do. Knowing this, it can be tempting to search around for these great managers, ignoring those with heavy processes or without the 'secret knowledge'. This is a mistake. Management is a skill like any other, and in order to progress to the highest level of competency you need to start with the basics. New managers need process and checklists to help them foresee potential problems that more experienced managers would intuitively 'just spot'. It is only by thinking through everything that a senior manager would 'just spot' that a trainee can build up the knowledge to help them to manage more intuitively.

This chapter contains some tips for management that I've found useful, and covers the following ideas.

- **Get your incentives right**
  How do you deal with unhelpful ideas in order to make sure people keep coming up with good ones?

- **Build teams**
  Teams don't just spring into being – put the effort into cementing them.

- **Communication is key**
  Communicate, communicate, communicate!

- **How to manage: set goals, monitor, feedback**
  The three essential steps of management.

- **Teach risk management**
  Risk management is an extremely complex and advanced skill that can't be solved by process alone.

- **When working hard isn't the answer**
  Most problems are not best solved by cancelling your evening engagement and staying at work until 2am.

- **Explain the commercial reality**
  It will be easier to incentivise the right behaviour if the underlying commercial picture is clear.

- **Is burnout a management problem?**
  Don't let your people wear themselves out.

Get your incentives right

Sometimes people come up with stupid ideas. They want to develop a product that is clearly never going to make any money. They want to spend half a day of everyone's precious time rearranging the furniture to make it more relaxing. They want to rewrite an obsolete codebase that no-one's even using anymore. They also want to chat about things – they disagree on the new policy, they wonder whether a few tweaks wouldn't make it more effective, they don't like one of the people on their team.

You're busy. It's very tempting to simply dismiss these ideas and requests – especially when it's always the same peeps who come and bug you. Why can't they be more like Shanice, who just gets on with her coding?

But if you dismiss them without considering their ideas and giving honest feedback you'll soon find yourself with a worse problem. You won't be getting people stepping up with good ideas. The thing with ideas is that there's a very fine line between a great idea and a blooper. It can be difficult to tell the difference; sometimes, you're going to get it wrong. But if people don't think their ideas are valued, they'll stop bringing them to you.

I have a simple set of rules for dealing with ideas.

1. Does it cost anything? Is there any risk of it causing problems? If there's a way we can trial it without a large cost or risk, then I'll sign it off as a trial. Continuation past the trial end-period will depend on whether the benefits promised were realised as expected.

2. Either way, give honest feedback on what you think of the idea. Is it great? Are you sceptical but willing to give it a go? Is it too expensive or risky?

3. Be enthusiastic about the fact that you've been brought an idea. Where it doesn't conflict with the point above, be enthusiastic about the idea too. Remember, this is how progress is made!

I also have advice for developers who find that their ideas keep getting batted back. To increase the idea of getting your idea adopted, make sure you include two key points:

1. Why is it not going to create any work or cost any money for
   the person who needs to sign it off?

2. How are you going to ensure that it doesn't cause any problems or
   repercussions for the person who needs to sign it off?

Build teams

In his project-management epic *Peopleware*, Tom de Marco talks
about the magic that happens when teams 'gel'. When a group of
people is functioning as a unit, they produce better results faster,
and, almost more importantly, they all have a lot more fun.

It's important not to forget that you want your whole tech department to work as a team. Shared office space and team meetings aren't
enough to turn a group of people into a team; they also need to
have fun together. Some particularly charismatic and energetic individuals can make work feel like play – they can weave threads
around people so that they're acting as a team before they know it.
But as a leader of a tech team, you can't rely on someone like that
being around – such individuals are, after all, few and far between.
Although the goal is for everyone to work together and have fun
while they do so, you'll probably need to start outside the office and
work inwards, rather than the other way around.

You need to make sure any out-of-hours socials don't feel like extra
unpaid work. That means no team-building exercises, no speeches
and no undercover work agenda. It has to be genuine fun. To achieve
this, you're going to need to do some work to find out what your team
actually does find fun. Paintballing can be one person's challenging
sport and another's cold, wet misery. I've seen team events ranging
from BMX biking to karaoke. It all depends on the participants.

Early stage companies populated by young graduates tend to cele-
brate and bond by visiting the pub. Companies with non-drinkers,

parents, remote workers and employees with other commitments need diversified events accordingly. As they grow, companies need to move away from 'evening down the pub' to shared cooked lunches, afternoon teas, and drinks in the office at 5:30pm to allow those who need to get off to have a quick chat and not get home too late. Offsite teams can gel together if given their own mini-budgets with which to arrange their own events.

Communication is key

I have always believed in distributed decision making and freedom to experiment. Even so, there are sometimes messages from the senior management team that need to be communicated.

Common messages include notes on employees' responsibilities or information about resources that could be used to help their personal development. Less frequently, there'll be instructions for company-wide processes that are taking place (e.g. appraisals and development records).

Misunderstanding of process isn't a problem. An employee can either ask, find out, and make the correction, or the misunderstanding will be discovered in checks and balances and be rectified. But even leaving aside wilful misunderstanding, techies can often get the wrong end of the stick regarding softer guidelines and recommendations.

There will always be some people who misunderstand what you say, no matter how much time you spend planning your speech and ironing out the ambiguities. To get the best results, make sure you give your message in as many ways as possible – and get your managers to follow up.

Another common mistake with communicating difficult messages is to not communicate them at all, or to try and soften their impact by emphasising the benefits and omitting to discuss the disadvantages.

The problem with the first approach (i.e. to simply not disclose things) is that information leaks. What's more, if you're worrying about something in your business (difficulties with sales, for example) and you don't let on that you're thinking about it, you end up giving a much worse impression! 'Why is the management talking about the new table football table when they should clearly be worrying about the sales situation?' would be a common question.

A similar objection applies to the second approach. If you're not talking about the elephant in the room, developers will think that you're trying to mislead them, that you're plain stupid, or both (you hired them because they were clever, remember). Worse, if you become perceived as someone who only communicates via spin, whenever you launch a new initiative everyone will immediately start looking for what you're not telling them, and very possibly starting finding issues where there aren't any.

How to manage

1. Set goals

There are three main steps in managing technical people. As we've seen, you're not going to be able to micro-manage them, for several reasons. Even if you were technical once, you don't have enough time as a manager to stay at the cutting edge, so it's unlikely that you know what they need to do. Secondly, you don't have *time* to micro-manage – you've got a whole load of extra reporting and meeting people that's not going to do itself, and you can't spend all your time working out the details of each developer's tasks. Finally, you need them to be stepping up and using their brains – it's how you get the best value from them.

The alternative to micro-managing is to set high-level goals for output, possibly alongside a vision. These two concepts overlap slightly, but essentially by 'goal' I mean a description of a task or other discrete output that needs to be produced, whereas by 'vision' I'm talking about an overarching ethos or way of producing the output. A simplistic example would be the goal of building a shopping application, which would be coupled with a vision of excellent UX and alignment with business requirements.

Don't forget that for a goal to be meaningful it needs to have a date specified. Development tasks will usually also have task estimates (for example, 'this task will be successful if it takes 10 days or less'), but these are really just a variation on end dates.

It's very tempting to make your own 'correct' goals for your team. If you used to be a coder, you probably know how long it would have taken you, and you don't want to be giving the people beneath you an easy ride. If you're not technical, it will be even worse – I have heard many cries of 'It can't possibly take that long!' But rather than being motivating, tough, one-sided goals give the people responsible for delivering them a get-out clause. The goals were unrealistic, everyone knew they were unrealistic, the business requirements changed, someone else let the team down. For a goal to be an effective motivator, the person who is delivering the goal needs to be signed up to it. The best way to get someone bought in is for them to set their goals themselves.

A lot of managers will worry that their teams will set themselves goals that are too easy. Funnily enough, teams often do the exact opposite: they set goals that are too hard. And these are even worse! If you know a goal is too hard, how can it be motivating? If a team knows that whatever they do they won't hit the goal, how can they work towards it?

Until your team is well-trained in setting goals, it may not be sensible to just leave them to get on with it. Setting goals that are achievable and challenging is very difficult. It requires a lot of understanding of exactly what is possible, and it includes (however you dress it up) the necessity of having to estimate pieces of work. You can do it via Scrum story points (see the illustrative example on p99), or estimate in hours or days, but you do need an idea of how much work there is.

The good news is that setting and hitting goals is quite addictive. Once you get into a pattern of successfully achieving what you set out to, you start to feel like a winner – especially if other teams are often late or behind on their goals. Teams with a strong goal-hitting culture will not only deliver more reliably ('one feature for sure' is almost always better than 'two features, maybe'), but will also start to increase the difficulty of their own goals as the ones that they are hitting get too easy.

Of course, your job also requires you to pass on requirements from the business. Sometimes these are the most effective goals. The business needs a new feature two weeks on Friday, otherwise we won't be ready to launch. The business needs this extra screen, but only if it can be developed within 10 days – can you do it? This communication must work both ways. The incentive for the business is to get as much done as cheaply as possible; left to itself, the business will stipulate unachievable goals *every single time*. The technical team needs to feed back to the business in this case, before the business makes commitments based on deliveries that simply aren't going to happen.

Agile development also has a concept of short-term goal-setting. All development is done in 'iterations', regular time periods, usually between one and four weeks long, in which the most important subset of the total functionality is developed.

It's important with goals to explain what is required and to gain consensus through good understanding. Goals are not a target that can be 'gamed' by leaving out other factors that are essential to the goal. For example, if you agree a goal of delivering a piece of software in a certain number of days, but then achieve it by omitting unit tests, or by delivering the feature containing bugs or omissions, the goal has clearly not been met. Any goal is therefore a communication from one party to another about what, in the most holistic sense, is required, and by what date that can be achieved.

A good example of how successful goal-setting can turn a project around was a team working on integration with a payment provider. The project started slowly, with framework setup eating up more time than expected. The first stories were completed much more slowly than they should have been. There was a sense of fatalism on the team; they had a number of story points to complete each week, but they knew that it wasn't going to happen so they didn't even try. The project was behind, but no-one seemed to have a sense of urgency about it.

The first step to recovery was examining *why* the goals weren't being hit. The most obvious reason was the one that everyone knew: the framework setup was taking too long, and the early stories were having to absorb some costs that would pay off later in the project. But by how much? By unpacking the framework points into separate stories and re-estimating accordingly, the project manager was able to communicate what was needed for the project to be successful, and the team immediately worked with more conviction towards the realistic (although still pretty tough) goal. They became one of the best-delivering teams, and turned the project round from a disaster to a triumph.

Vision is a slightly looser equivalent to goals. Only sophisticated and well-functioning teams can be led by vision. In high-level terms, as a team, what you need to deliver is always going to be the same: high-quality code at the best value for money. There may be additional requirements in your team, or different balances between quality and productivity, which will combine to produce an overall vision. If your team is well-versed in setting their own goals and delivering in accordance with the vision, you need only communicate that vision well and sit back and watch the team deliver.

I once worked on a long-lived product for a digital media client. The software ran across several mobile and desktop platforms and had a very sophisticated feature-set developed over five years. At any one point, there were many things that need to be managed: support requests on the existing platforms, development on new platforms, underlying maintenance of the code, and a long backlog of new features and enhancements that would allow the client's sales team to sell more products.

The client provided us with a product owner. The product owner met with the technical team at least once a week and explained how the business was prioritising different development efforts. On top of that, the team had been set a vision: the product was the core of the client's business, and needed to be both robust and scalable to keep existing users, as well as feature-rich in order to expand to new users. This vision was well understood by the team, and meant that, rather than the client product owner setting us detailed goals, the team could (and did) challenge the immediate prioritisation of work when it didn't conform to the overriding vision that they have been set. This ability to 'work to the vision' resulted in, for example, critical server availability work being carried out in preference to more exciting new development. This server availability work paid off in spades when a server fault took an entire disk array out. If you have a team of highly-skilled technical staff, make sure they understand the bigger picture to which they need to deliver. This will allow them to deliver the best results.

Illustrative Example: Story Point Estimating

Estimating software is hard. It's notoriously easy to end up with estimates that are tens of times out. The standard way to deal with this is to break estimates down into smaller tasks, and to do a large amount of investigative work in order to improve estimate accuracy.

An alternative approach is taken when estimating Agile stories (tasks) using the Scrum methodology. Instead of estimating using a time period – say hours or days – to estimate a task, an abstract points system is used. The size of a point is consistent across stories, but doesn't refer to a particular time period. The rationale behind this kind of estimating is that at a high level it is much easier to estimate relative size than to estimate an exact amount of time for delivery. A set of features that is very clearly larger than another can be estimated with a higher order of accuracy as 'double the size' or 'five times the size' than as a certain amount of work. This also helps to mitigate developers' tendency to under-estimate the amount of work required for a story in an attempt to please the business (in the short term; of course, in the long term, the business will be more unhappy rather than less).

When estimating approximately, it's often cost-effective to involve the whole team, in a way that it really isn't if you are estimating by drilling down to the detail. A simple way to get a varied input rather than relying on a single estimator is to get the whole team in a room and give each member a set of cards bearing the Fibonacci numbers (1, 2, 3, 5, 8, 13…). For each task, every member of the team holds up the card that they think best corresponds to the relative size of the task. A suitable average can then be taken to get a definitive points score for the story.

Once a certain number of stories have been completed, a 'velocity' can be calculated by looking at how many points are delivered within a certain time period. These can then be used to determine how many points of work can be completed in the next time-boxed period. Unfortunately, once you have calculated a velocity you lose the benefit of time-independent points, as the team will now be very aware that a point is calculated as taking a day or 1.5 days. This may influence their estimating accordingly.

2. Monitor

If you've been managing a team for a while, you may have found the concept of managing a team through nothing more than a 'vision' somewhat alarming. Just because some goals have been set, it doesn't mean that they will be hit; it takes a lot of successful development projects to earn enough trust for a team to be left to manage themselves.

In fact, any team good enough to be 'left alone' will be employing the same techniques that you need to use with new or unfamiliar teams in order to understand whether they are on track – and, if not, what they need to change.

The key is to monitor key incremental indicators and extrapolate from these. While a vision can be high-level and in many senses qualitative, monitoring must be numeric. Essentially, visions must be distilled into concrete goals, and concrete goals into key metrics.

There seems to be a mismatch here. How can a vision such as 'The end user will delight in using this app' be translated into a numeric measure? One way to make sense of this is to ask yourself the question 'How will I know if this is successful?' How will you know if the user is delighting in the experience? How will you know if the product delivered good value for money? How will you know if the project was on time and within budget?

Either there's evidence that something is the case, or there's evidence that it isn't (or, a third option, it can't be observed – in which case, managing the team is going to be least of your worries; you simply won't be able to know whether you've delivered it or not). If you're given requirements from the business or an external client, insist that they answer the question 'How will you know if the system is delivering as you required?' You may need to help them out and suggest some metrics, but unless you're sure that the client is expecting the same outputs as you, you may have a nasty surprise when it comes to User Acceptance Testing. If you have an excellent business analyst on the team or have come to understand the business requirements well yourself, you may be in the exalted position of being able to take the business vision in simple English and translate it yourself into metrics that you know will deliver on that vision. Either way, you need to have these incremental metrics available in order to be able to see whether the project is on track.

When developing software, the most obvious way to break down a project is by unit of development work. This may be based on an abstract points system or on estimates in time such as days or hours. There will also be attendant requirements that need to be delivered on top of feature development, such as performance and other non-functional requirements, testing, and management activities. It's a matter of taste whether or not you assign separate estimates to these and track how long they will take; however, a Scrum burnup chart is a great way to strip a project down to just the important information.

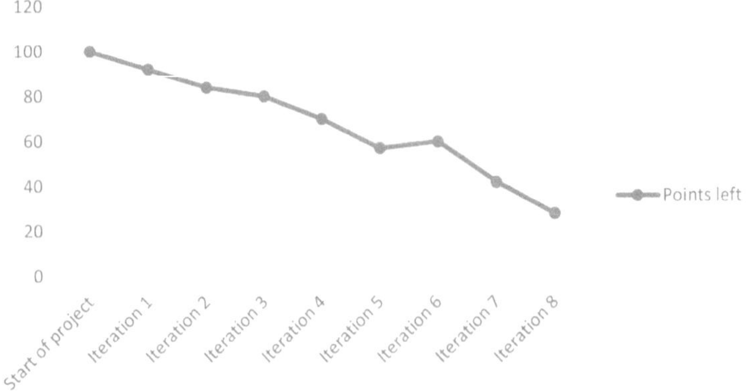

If you practice Scrum yourself you might be more familiar with a burn*down* chart. I'll explain that first. You count your total estimated units and mark those on the graph; then, each week, you count the number of units left and mark those on your graph. You aim for a nice smooth line down towards zero, at which point the project is completed.

For the purposes of a burn down chart, only development points are counted towards the total points. It is assumed that other overheads are absorbed into the development in a roughly linear manner. It's important that development points are counted as completed only when they are either put live or completely ready to go live. One of the traditional pitfalls of software development is '90 per cent done syndrome', where a feature feels nearly ready, then some more work is found that needs to be done, then it needs to be tested, then some defects are fixed… (a great joke: 'The first 90 per cent of the code accounts for the first 90 per cent of the development time; the remaining 10 per cent accounts for the other 90 per cent.'). It's very tempting to try and count some points when a feature is *nearly* ready, but you'll achieve a much better result if you only look at truly completed features.

You'll see that the chart above achieves smooth burndown until we hit the end of Iteration 6. It appears that the team ended Iteration 6 with more points remaining than when they started. Did they trash the codebase?

A Scrum burnup chart

A burnup chart helps us to see more clearly what has happened here. A central tenet of modern software development is that change is to be expected and accepted rather than feared. Theoretically, change could result in either more or less development work. In practice it rarely results in less.

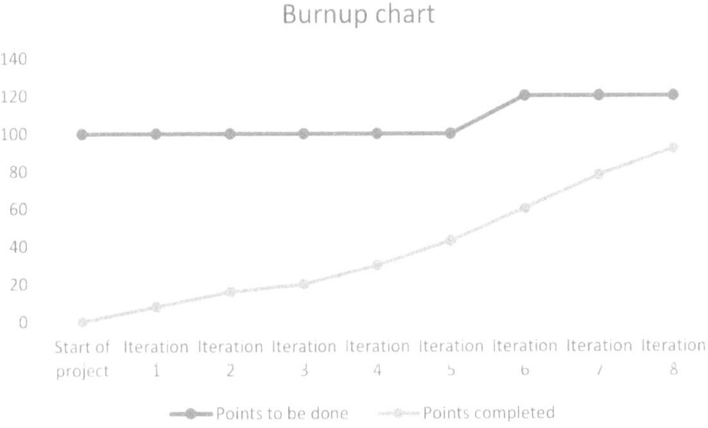

In a burnup chart, you count the total number of development points to do, and track them in one line of the graph. You then track how many points are done – but instead of deducting these from the number required to get a graph of the remaining time, you simply plot the number of points done.

As you can see, during Iteration 6 extra work was added to the backlog. The team were delivering work at the same rate, but the target had been changed. The burnup graph helps you to separate out 'work done' from 'increase in scope' and avoid the demotivating feeling of going backwards or slowing down due to scope creep.

Of course, you'll already be aware that 'how many features' are delivered is the aspect of software development that is most readily measured via metrics. But how do you measure whether the code being produced is of the right quality?

Step one, which it is unforgiveable to miss out in modern tech development, is a suitable amount of automated testing. Automated testing catches embarrassing errors and prevents them from making it to test – or, worse, into a live release. With estimates of the proportion of total development effort incurred post-release stretching up to 80 or 90 per cent, you need to take the long-term 'win' of building-in automated tests that will prove their value over and over again.

Years ago, I was slightly sceptical about automated tests. The argument that convinced me of their value came from Robert C. Martin in his book *The Clean Coder*. If you have good automated tests, you can have the confidence to change your code. You can refactor and improve performance without worrying about whether you are breaking an edge case or introducing subtle bugs. This is a very powerful ability.

Automated tests will tell you whether your code is working correctly, which is one of the most obvious quality angles. With the right tests set up, they will also tell you whether you are meeting your non-functional requirements: performance under load, speed of execution, etc. But they can't tell you how maintainable and well-architected the code is.

This may sound trite, but the first thing that you need in order to be able to tell whether code is well constructed is someone who understands exactly what this means. If you were a hotshot developer yourself, you can skip the next few paragraphs; the chances are, however, that you were not, or if you were it was quite a while ago, and you're not able to cast your eyes over the code yourself to determine whether it is any good or not.

This is the point at which you need to start learning from your developers (see *Who's the expert?* p57). In fact, unless you're 100 per cent sure that your lead developer knows all the ins and outs of the matter, you may need to learn from external sources too. In order to verify that code has been well developed, you don't need to know how to code it yourself, but you do need to have an understanding of the general principles and be able to have a discussion about them.

Don't ask questions that have easy get-out clauses. It's hard to answer 'Have you followed industry best-practice guidelines?' with anything other than 'Yes', particularly if you're pretty happy with your code and would quite like to get off home rather than justify your work to someone who doesn't understand anyway. If you can present challenges such as 'Show me how large your methods are and how they are easily maintainable' and listen carefully to the answers, you'll not only be able to ascertain much more clearly the quality of your codebase but will learn something, too. Industry standard books such as *Clean Code* and *The Pragmatic Programmer* are good for picking up this background knowledge, as are language-specific books.

But remember that good code quality is not merely about following a set of rules. It's not necessary for your developers to have followed the latest coding fad in order for their code to be well constructed – in fact, coding too much according to 'the rules' is usually a very bad sign (remember that being a developer is all about using your brain – see p54). I'm not suggesting that you read up on the latest trends and then harangue your developers about why they are not following them; you'll lose their patience very quickly if you try that. Instead you need to be able to ask questions and engage in discussion to see whether they are considering the right kinds of things. You can also follow external cues, such as whether they are open and constructive when they realise that a mistake has been made, whether they honestly consider all the options, and whether they are interested in finding out about new technology trends themselves and working them into their skillsets.

Once you have a certain amount of faith that your team are going about things the right way, code quality is a responsibility that they can take on themselves. As dedicated professionals, they will be interested in reviewing each others' work and holding the team to the highest standard. Personally, I would still take an interest (leading questions such as 'Have you had enough time to get the codebase to the quality that you would be proud of?' offer a nice get-out, allowing the developer to blame something else, in this case time pressure, and so admit to lower standards of quality), but as a now reasonably long-time ex-developer I know that I am not capable of assessing code as well as senior technical people.

It is generally a good idea to practise peer review. That way each piece of code committed to the code base has been seen by at least two sets of eyes. But peer review has another benefit besides code quality. It's one of the best ways for developers to learn good coding practices. Each time they write a piece of code, they have the opportunity to ask a developer with more experience how else they could have approached it. This way, knowledge of tools and libraries is passed on by osmosis and ideas about how to implement common structures discovered though discussion. This provides an incentive for developers to ask the strongest available developer to review their code, and so naturally encourages a strong codebase.

So peer review is a great tool. Sometimes, however, if a team becomes too insular, they can accept certain aspects of the code as necessary or unchangeable. Adding in an extra review step several times throughout the development project by a senior member of *another* team helps to share knowledge between teams and shine light into dark corners.

I won't go into detail on other metrics that you may need to measure, such as adherence to the specification or the unbridled joy of the end users – but I will set out some general principles for measuring such metrics incrementally.

Anything that you can judge at the end of the project, you can judge now. If you want to know how users will interact with the system, you can create paper or HTML mock-ups and test their interactions. You can do this work before you start coding (this is almost always the right way round) and then code to what you already know. And don't forget to release at points throughout development so that users can look at the system 'for real' and test their interaction.

Do you know that users will change their minds? That they'll like one feature one day and change their minds the next? Don't throw your hands up in horror at the users. Instead, work out a plan for how to deal with this, and how to judge whether your system is a success despite this complication. Remember that other people will be judging the success of the project too, and the sooner you agree the criteria that will be used the easier it will be for you and your team to come out victorious in the end.

Adherence to the specification does perhaps warrant a quick note. Testing was originally designed as a verification step to check what had been developed. Now a lot of testing work – such as working out exactly what needs to be verified – fits much better at the start of a project. This is the subtle difference between 'Does it do X?' and 'What is the X that it should do?' Working this out at the start of the project not only saves time at the end, when there is traditionally an inelegant scramble, but also allows developers to have a clear direction in which to develop and can significantly reduce the amount of re-work required.

3. Feedback

Setting goals and monitoring are the steps that managers tend not to overlook. After all, they have to report upwards to other managers, clients and users, and they will be asked for this information themselves.

What *can* be forgotten is that what is obvious to you is not obvious to your team. Clever as they are, your team can't read your mind. If you don't provide clear feedback on what's being done well and what's being done not so well, your team can't adapt their behaviour to achieve what you need.

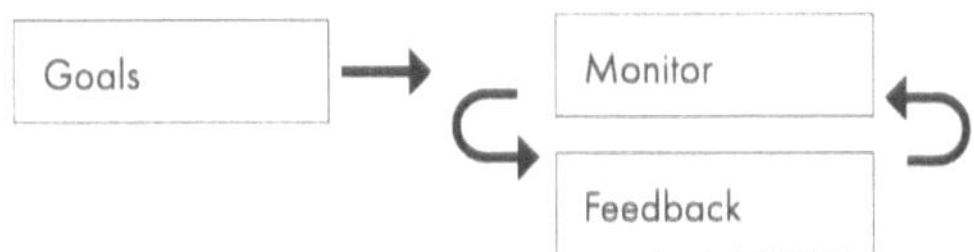

Giving direct and honest feedback is one of the most difficult skills that a human being can master. It is also essential to the management process – particularly so with technical people.

As we have seen, technical people are creative and innovative, and not, as some people believe, as mechanical and computational as the machines that they control. The same is true of their emotional responses. Technical people are often quiet and not prone to open and violent histrionics, but this doesn't mean that they aren't emotionally perceptive and won't respond emotionally to positive and negative feedback.

I first started to put together a cohesive picture of this issue after reading a book by the Dale Carnegie Foundation. The book was called *5 Essential People Skills*; essential people skills sounded like exactly the kind of thing that I needed to learn. It had some good ideas, which I dutifully noted down, but it was only on investigating further to find out what other titles this mysterious 'foundation' had produced that I realised that its books were based around the work of a single author, Dale Carnegie, and that the foundation seemed to have been set up solely to trade on his name. Clearly, he must have established a very powerful name.

I had heard of Carnegie's book *How To Win Friends And Influence People*. Going by the title alone, I had considered it to be a book for political machinators, or people looking to get ahead in business in an unsavoury way. But upon learning that it had evidently inspired some pretty sensible business books, I decided I should take a look.

How To Win Friends And Influence People was written in 1936 and is one of those amazing books that seems to be completely relevant to today in spite of its age. Aside from a few dubious references to touching people on the arm (it's my understanding that this is now universally considered pretty creepy – at any rate don't try it on me), the book reads as an accurate description of the business and personal situations that we encounter every day, along with advice that you can see will help you communicate with other people. The reason for this is quite clear. Dale Carnegie is writing about people, and how people behave. While the environment in which people operate has changed (computers being the least of it), people them- selves have not.

Building on what I've learned from Carnegie's book and others I have since read, I've formulated a few basic rules for giving feedback.

1. Work out what feedback you need to give and be direct.
I think the first and most obvious pitfall with giving feedback is to shy away from what you're really trying to say. I'm not advocating that you 'let rip', or say exactly what you would say in the heat of

the moment, but you do need to work out exactly what you are concerned about and put this clearly to the person you're feeding back to. If you mean 'You are working too slowly', don't say 'Are you sure you're getting all the support you need?' Human interaction is already fraught with ambiguity, where one side understands one thing and the other side another. Every time you use a euphemism you are increasing the chances of you thinking you're communicating one message and your audience hearing something different – or even the opposite.

2. Be specific.

This is a great tip that I regularly struggle to apply. I like to be aware of it because it is one of those things that is difficult and forces you to realise that you're not thinking your feedback through clearly. The nicest piece of feedback to give someone is 'Great job!' It makes them feel good, and it makes you feel good too. But *why* was it a great job? What made it better than the other jobs that didn't get the same praise? It can be hard to work this out, but it is absolutely essential to do so if you're going to get the full benefit of feedback and inspire the person that you're feeding back to in order to emulate the same behaviour again. A lot of different actions can go into an end result; unless you're specific about which one you're praising, the wrong one might get repeated instead.

This is where it gets tricky. Often my subconscious brain seems to know better than my conscious brain when someone's done a good job. I find it can be really easy to say 'Great job!' – and to mean it – yet hard to break it down into exactly what was done well. Was it that it was delivered on time? That it was accurate? That the job itself was done to the usual standard but was done with exceptional good grace and a smile?

Specific feedback is more believable. If someone praises a specific thing that I did or a specific way that I did it, I know that they were paying attention to what I did. I can see that they're not just rolling out the same platitude to every single person who does a job for them.

Finally, a smaller point is that specific feedback also explains to the people that you are not feeding back to why you chose to single someone other than them out for praise. It may encourage the right behaviour in these other members of your team too, because now everyone knows what you are looking for.

3. Criticise the behaviour not the person.

Our language when we need to feedback to someone tends to link how they have performed with some unchanging aspect of their personality: 'You're a great developer', 'You never test your code well enough' or 'You're always late'. But we obviously don't believe that. Deep down we must believe that it's possible for people to change, otherwise we wouldn't bother feeding back to them at all.

Using language that implies that failings (or, equally, good performance) are a permanent part of someone is not a great incentive for them to change. If I'm always late, why should I bother trying to be on time? If I'm always criticised however I try to test my code, why try to be any more thorough?

It's important to decouple the bad action from the person who performed it: 'This month's report was late Jane – that's unlike you'. If you can refer back to better performance, do so, and encourage comparison between the two: 'You didn't test this piece of code well enough. You did a much better job on the project last week – what was different?'

And of course, once you do start to see improvement, make sure that you notice it and highlight it. Where possible, break improvements down so that they can be resolved sequentially, reinforcing the idea that behaviour can change for the better.

4. If you're upset, wait and calm down.

When you're let down, you react emotionally. You have your own goals and targets to deliver, and maybe your boss hasn't read any books on how to give feedback and encourage people, so you're not going to be able to have a nice conversation about how you could have done things differently – you're going to be faced with raised voices and waved hands. So when you're forced to miss a key deadline that you have worked hard for because someone in your team forgot something, didn't take enough care or went home early when they were needed, you're going to be upset. That's only natural. But this is not a good time to feedback to a person who needs to improve their behaviour.

If you're a slightly reticent manager, you might find that this is the time that you give the most feedback. If you're not upset, it might not feel worth having a difficult conversation. If someone's habitually late and it doesn't matter, you might be tempted to avoid making a scene and let it pass. But the one day that they're late and you really needed them to be on time, you'll fly off the handle.

We all know that when we get very upset with someone, for example in row with a family member, we both overstate the behaviour that we're upset about and allow it to bleed over into other areas that have nothing to do with what we were upset about in the first place. If you're trying to give a clear message about what behaviour is useful and constructive and what isn't, this is not the right state of mind in which to try and do it. You won't get your message across clearly, you may overstep the mark and make personal comments, and, worst of all, anything you say may be dismissed as something that you were only saying because you were upset.

5. Make improvement seem easy.

Often you'll be feeding back on behaviour that seems really obvious to you. You asked for work to be delivered on time, you asked to be notified if something was going to be late, you asked for clear and regular status emails, or something similar. It was simple. What possible reason could there have been for someone to not deliver it, except that they just didn't bother?

You might think something is simple; the person that you're asking might not. In fact, surely it's clear that they *didn't* find it as easy as you did – otherwise they would have done it.

If someone has a chronic problem – they seem unable to deliver work on time, say – you will often need to do some work to find out the underlying problem. Are they taking on other tasks to help out other teams? Are they failing to prioritise their tasks? Remember, if you're giving feedback in order to encourage someone to change, you need to have a clear plan for how they can do this. Otherwise they might end up being keen to improve but trapped in the same patterns of behaviour.

6. Positive feedback first, constructive feedback second.
A great book on how to manage is *The One Minute Manager* by Ken Blanchard – which beats many other books solely on the criterion of how long it takes to read. *The One Minute Manager* outlines three key ways to manage, in a set order.

First you must set a one-minute goal. This is a clear brief for what you are looking for your managee to achieve (we cover this in the section *Set Goals* p94).

Second, you must give one-minute praisings. Following the feedback rules we've talked about in this section, you should give specific praise whenever something is done well. Until you've built rapport with your managees you are not allowed to move on to the next stage.

As human beings we tend to assume that everything will work as we expect, and only notice when things aren't going to plan. This also holds true where we, as managers, expect people working diligently and focussing on the right goals to be the normal state of affairs. We notice when something is late, or has mistakes in it. We don't notice when something is on time or correct.

The One Minute Manager's key message is that you need to change your behaviour as a manager to see when things are going well. Appreciate the work that your team is doing for you. Praise a great piece of code, a successful delivery, a great attitude or a willingness to help others. Don't wait for things to go wrong before identifying the difference between good and bad work – identify the good work straight away. As Blanchard says, 'walk around the office trying to catch people doing something right' – go and look for people doing things well, and point it out to them.

Having a manager who recognises the things that you do well and the hard work that you put in builds trust in a relationship and a sense that they are on your side. It is absolutely essential to have this in place before you get to Blanchard's final tip for management – the one-minute reprimand.

This whole section is about giving clear and direct feedback, and there is simply no way that you can manage a team well without having to point out when people could have done better (I would probably avoid the word 'reprimand' myself). But it's important to realise that you need to create an environment where your feedback is going to be interpreted as helpful and constructive – otherwise it can be rationalised away.

Some management books suggest a ratio of as many as five pieces of positive feedback to every piece of negative feedback. I've never managed to achieve this myself, but I find it a useful target to aim for.

7. Encourage two-way feedback.
Another way to build an atmosphere of trust is to ask for and act on upwards feedback. Lots of organisations now practice '360-degree'

feedback, where everyone feeds back on everyone, but getting your team to complete an anodyne form is not going to help much. You need to be keen to understand what they think and willing to accept that you might be wrong. You're asking it of them, so doing it yourself seems a good way to model the behaviour that you're seeking.

You'll need to be a strong person to genuinely engage with this. If you're worried that your team might expose any weaknesses you have, or that they don't like you, or that the criticisms they have might filter up and damage your reputation elsewhere, it's not going to work as a technique. But remember that, however difficult you find this, they will be feeling exactly the same, especially as you are most likely the one filling in their official performance reviews at the end of the year.

8. Train people to find their own feedback.

The easiest people to help are those who analyse their own performance and consider dispassionately whether they could have done something better. It's so common to interpret the question 'Could I have done better?' to mean 'Ought I to have done better?' Analysing performance then becomes a blame game, where you are keen to show that things are down to others' failures while your own performance has been flawless. This is the worst environment for learning and improving.

If instead you can help people to understand that there is always something that *could* have been done better and that how you deal with this is more important than not having performed perfectly, you can encourage them to explore things that might have helped with the project.

One of my most self-aware managers came up with a thought experiment to illustrate this perfectly. Imagine that every other human being in the world was a robot. They respond in the same way as regular people – they are deterministic, but not predictable – but they have no independent will. Now imagine you have to deliver a piece of software in this environment. Quite clearly, none of the other 'human beings' can be blamed for anything – they are simply responding to stimuli. This thought experiment can help to illustrate very clearly just how much can be changed just by your actions. It can also take away the folly of blaming yourself. After all, if everyone else is a robot, they won't be blaming you.

An Agile retro is a team example of this approach to learning. The team retro is not designed to assign blame to the individual who broke the build or to pick on the isolated piece of code that isn't up to scratch. Instead the team looks at the bigger picture and the higher-level processes in order to identify improvements. This mechanism of the whole team taking responsibility can also be a useful way of avoiding unproductive blame, encouraging fair and objective analysis, and achieving genuine improvement.

Teach risk management

To a project manager, risk management is the ultimate tool for ensuring project delivery. Good risk management is inherent in all project-management styles (if there were no risks to manage, there would be no challenge), but as a technique and a discipline it helps project managers to take decisions that they might otherwise have overlooked.

This is an important principle to remember. The risk log (or risk register) is a tool. It's not a complete repository of all risks that could ever impact on your project. This simply isn't possible, for one thing, but even if it were, a risk log containing all risks that could ever impact on your project would be unmanageable. It would take weeks just to read through it. You would never be able to action any mitigation steps, or even get any other work done!

Formal risk management can often miss this point. In formal risk management, the following goals generally apply.

1. Show your manager that you wrote a risk log.

2. Make sure you followed the standard layout so that your manager can tell that you did it 'properly'.

3. Make sure that there are a sensible number of risks on the list.

4. Include the kind of risks that you know your manager will want to see on there; in other words make the risk list look good to a quick skim by someone busy and not directly involved in your project.

In listing these steps, I'm not trying to criticise or belittle risk managers. I'm just trying to point out the kind of pressures that usually exist in work environments and how people respond to them.

A formal risk process will reassure a company that risk is being taken, and managed, seriously. In fact, as I learn about new risk-management techniques (such as transition indicators, of which more later), my instinctive reaction is to include them in a template so that junior project managers can start managing risks well with no extra ramp-up or learning.

But you'll already know that this is not how knowledge gets transferred. I can't train a developer by writing out a list of simple instructions or getting them to fill in a table with checkboxes. The reason that a risk format doesn't solve the problem on its own is the same: you need to know what to put in the boxes. Understanding exactly what a heading means and how you can optimally use this to control risky aspects of your project is not straightforward. Filling in the boxes 'Project name' and 'Major risks' are not equivalent kinds of form filling; the latter is the record of the culmination of a thought process.

What's more, a form-filling approach to risk management can engender a worse result than no official risk-management process at all. This is for the simple reason that any kind of process is designed to set your mind at ease. Take my task-management process as an example. My task-management process is designed to ensure that, by careful management of the master list (a post in my Outlook inbox), I ensure that everything that I need to remember to do gets on to the list, and hence that I'm reminded of its existence when I get to an appropriate day for dealing with it. I don't need to remember any deadlines or tasks (and you'd need to be a memory master to keep my task list in your head at any one time). A risk log is a similar tracking mechanism for risks.

There are two ways to manage a project. The first is by holding all the necessary information in your head, reacting to change as it happens, thinking ahead, seeing potential upcoming obstacles and sweeping them out of the way – in short, by living the project. Having seen other ways to manage, I personally find this approach to be tiring and stressful. It's also subject to human error. I don't like to be in the position where someone asks 'Why wasn't this done?' and all I can say is that I forgot.

Hence the second way. This way, you write down the risks, along with the dates that you need to deal with them, so that you don't have to be constantly thinking about whether anything needs actioning. If you adopt this approach, which most managers do, you must remember that you have switched off the part of your brain that is worrying every minute of every day about the risks. You'll be able to get to sleep at night, but if your risk log isn't up to scratch you'll quickly find that the risks you failed to consider start to materialise – and take a lot of work to clear up.

So, bearing in mind all these caveats, here are what I've found through practical experience to be the most useful aspects of risk management. Without exception I have worked these out by going through old risk logs, horrified at how they were not up to date, inaccurate, tracking the wrong kind of risks, or not forcing me to take the steps I needed to manage the risks.

This isn't a step-by-step guide to managing a risk log. You'll get the best results from working out your own format, designed to support your end goal of identifying and mitigating important risks that might cause you problems if they catch you unawares. If you have project managers reporting to you who need to manage risks, it's best to teach them that they need to acquire a good practical understanding of risk management and let them work out their own system. You might have worked out a system that helps you to stop at the right places and consider what needs to be done, but as soon as you turn the process into a form-filling exercise your project managers will not be getting the same benefits. Force them to work out what works for them.

1. Don't confuse risks with issues or actions.

This is the most important point, so I'm going to state it upfront. Risk is a synonym for *hazard*, the French word originally used to mean *probability*. Risks are probabilistic events. This is what distinguishes them from actions (things that you must do) or issues (events that have happened, or will happen with certainty, that you need to do something about). By necessity, risks concern things that will happen in the future – but not everything that happens in the future is a risk.

Consider this statement: 'Our burnup chart shows that there is a risk that if we don't change our behaviour we will not hit the delivery dates.'

The chart does not show a risk. It shows an issue. It shows that, for sure, if you don't change your behaviour you won't hit the dates. This requires a different response than a risk. An issue is actually much easier to deal with, because you know what you have to overcome. If you are in a situation where you don't know whether or not you are going to hit the deadline, the first step is to work this out which is the case. This is turning a risk into an issue, or resolving a risk into a certainty. This is not to be confused with the fact that, because estimates are never completely deterministic, there is a risk that the expected delivery date may at a later point transition over to the wrong side of the deadline. To manage this risk, you need to schedule regular checks (see Point 6).

Many people keep a separate 'issue log' in addition to a risk log in order to keep the two concepts clearly separate. I prefer to turn issues into actions, and put them straight into my 'to do' list.

2. Riskstorming

Riskstorming refers to a group meeting in which the team brainstorms potential risks. This is a great way of getting information from the whole team, and getting them bought into and trained up on risk management.

A riskstorm takes a significant amount of time. A good start is to simply consider every type of risk (including identifying and ruling out the 'wrong' types of risk). A riskstorm every week would be overkill, but scheduling another mid-project (or every few months on a longer project) can help to keep your risk-awareness fresh and catch risks that you might miss in a weekly check.

There are many ways to run a riskstorm; here's a simple one (using some techniques taken from Agile retros).

Give everyone taking part a book of A5 Post-it notes and a Sharpie pen. (Yes, I did just specify the type of pen to use there. One of our senior project managers once said that 'all good management techniques seem to come down to stationery'. I'm hoping it was tongue in cheek – but anyway, a Sharpie pen is both clear enough to be seen but not too thick to read.)

In a time-boxed period (we generally go for five minutes – it's tight, and forces people to think quickly), ask everyone to write down all the risks that they can think of. At this stage it's not important to assess the severity or likelihood of the risks; any risk that may affect the project is fair game.

Draw two axes on a whiteboard: low impact/severity to high impact/severity and low likelihood to high likelihood. We usually structure the board as left = low impact, right = high impact, top = high likelihood and bottom = low likelihood. I'll use this ordering in subsequent descriptions.

Once everyone has completed their personal list of risks, add them to the board. Where someone has duplicated someone else's risk, throw the Post-it away. You can have everyone add all their risks at once for time efficiency, or have everyone add one at a time in turn so that the last person doesn't feel inadequate when all their risks have already been posted to the board.

When sticking a risk on the board it should be assessed by the whole room for approximate likelihood and severity and placed accordingly. When all the risks are placed, plans can be formed to deal with the top-right risks first. Risks towards the bottom and left of the board may be judged insignificant and not worthy of mitigation (as discussed above, some risks are not worth the effort of mitigating them).

When you're done, photograph the board for evidence of the risk-storm, and translate the actionable risks into a risk log.

3. You only need to record risks that you can do
   something about.

The most fun part of a risk storm is thinking of all the possible, crazy things that could scupper the project. 'The earth might get hit by a meteorite'; 'A solar flare might wipe out all of our data and our backups simultaneously'. These are real risks, and ones where the impact on the project would be severe, but they're impossible to mitigate. If the whole of the Earth's population is wiped out in a freak accident, yes you have no project team but you also have no customer to care that there's no project team! If you allow yourself to get distracted with risks of this kind you'll have less time to spend worrying about the significant risks that you *can* make an impact on.

4. You only need to record risks that have a significant chance of happening.

Similarly, there are lots of risks that you can mitigate against, but that are vanishingly unlikely to happen. A meteor storm is disastrous, but there's not very much you can do about it. This is equivalent to ignoring the extreme left-most and bottom-most risks on the riskstorm, particularly the ones in the bottom-left corner.

5. Understand the difference between mitigation and contingency.

I sometimes get the words *mitigation* and *contingency* confused. I know that one is for things you do before a risk occurs and one for things that you do afterwards (for the avoidance of doubt: *mitigation* is the act of trying to prevent a risk occurring; *contingency* is a plan for how to deal with the risk once it has become an issue).

It's actually important not to mix the two concepts up. Sometimes it may be convenient to keep them both in the same 'Mitigation/ Contingency' column in your risk log (by which you mean 'what I plan to do about this'), but they will tend to be very different types of actions, requiring different types of thinking.

To work out mitigation actions you need to think 'how can I prevent this from happening?' You're thinking about scenarios where the risk doesn't materialise, and trying to make the reality match up to one of those scenarios. On the other hand, to work out contingency actions you need to own up to the fact that the risk may well materialise, and start to plan for what you will do if it does. It's important that you wear both hats and think through both mitigation and contingency, although one or the other may be more appropriate depending on the risk in question.

One further note on contingency. You'll notice that I often use the phrase 'contingency plan' rather than 'contingency'. A contingency action needs to be something you can do *now* that will help if the risk materialises. Working out a plan can be helpful, so that you don't need to stop at that point and think again. What's most useful, however, is anything that you can do right now that will save significant time later. The best example is setting up a duplicate server to keep on standby in the case of failure. This is a planning action that can turn hours of downtime into minutes or seconds. One of my most common review mark-ups to risk logs is where a contingency plan equates to 'Deal with it'. No project manager is going to forget to reactively deal with something once it's happened. You should save yourself the review time in your regular check, and leave that out of your log.

6. Transition indicators are essential.

I'd been a project manager for several years before I came across the concept of a transition indicator. I mention this risk-log feature here (when I miss out many others) because I think it's an essential component of good risk management and, given this, is less well-known than it should be. If you know what a transition indicator is, you can skip this point (and if you know what one is and don't think it's useful, send me an email and we can have an interesting debate about it).

A transition indicator is a statement attached to a risk – e.g. in another column in your register – that tells you how you will know when a risk has occurred. This may not seem particularly useful; many risks ('business is unhappy with the design'; 'performance is too slow') seem like they would be obvious when they occurred. This is true, but that misses the key point of a good transition indicator. A good transition indicator lets you know ahead of time when a risk is going to materialise.

The most obvious way in which we use a transition indicator (even if we don't call it by that name) is when we consult a burnup chart to find out whether we will complete all of our development tasks on time. We could wait until the delivery date to see whether the risk of overrun has materialised, at which point it would be obvious – but it would also be too late to do anything about it. If instead we can track our progress so that we know at the start of the first sprint whether the remaining development work will be completed in time, we have a lot more options for turning things around and changing the outcome.

I think this is a key point about risk management. I've said it before, but it's well worth repeating. The main aim of risk management is to either turn risks into certainties or nullify risks via mitigation and contingency planning (official risk processes include other options, such as transferring the risk to someone else, but that's more of an arse-covering exercise than a risk management one). The latter is much more difficult than the former.

So the key to risk management is not to prevent risks materialising, but to determine as early as possible whether they will occur or not – to turn a probabilistic situation into a deterministic one. This also helps to explain why people are often bad at managing risks: because we fear the outcome materialising, we don't want to resolve it (because as long it's unresolved we can delude ourselves that there's still a hope that it won't happen). By considering resolving – rather than preventing – risks as your key aim, you can help avoid this inbuilt tendency and deliver more projects on time and on budget.

Once you have a transition indicator for a risk (see previous point), you should aim to bring the point at which the transition indicator is triggered forward as much as possible. For most risks, this is likely to involve adding a new regular check of some description.

For example, let's take the risk that the business won't be happy with the user interface design. The key to establishing whether or not this is the case is to ask them as early as possible. First up, you can add in a UX design and review step, which might prevent you from writing a whole bunch of code that you didn't need, but even after this step you'll need to schedule a regular review of the application to see whether it's still meeting requirements. Your transition indicator then becomes 'at my regular review meeting, X happens'.

It's remarkable how often having a transition indicator results in an outcome of this sort. It's not so much that you write down the point at which you expect to know, but that, by considering risks in this way, you find creative answers to the question 'How early can I know?'.

Risk management is often described as a proactive activity, but with good proactive thinking up front you can turn a lot of the rest of your risk management into a *reactive* process. This is much easier to manage. Whereas your instinct upfront is to not resolve a risk because it may not resolve the way you'd like it to, your instinct once an issue is uncovered will be to jump on it and get it sorted out. This is why risk management, like all good management techniques, is so effective – it works with human instincts rather than against them.

Most people have a process for managing their risks that involves a weekly or daily check. Once you have a transition indicator that relies on a check ('performance tests don't pass', say) it's apparent how often you need to check that risk. You'll find that some risks need to be checked up on more often than others. You don't want to miss a risk that might have been triggered daily, but you also don't want to wade through weekly risk checks more often than you need to.

It's possible to maintain separate lists for different frequencies, but that leaves you open to the possibility that you check one and forget the other. I think a nice solution is to order risks by frequency to check, check to the end of the daily ones every day, and review the whole list every Monday.

How often to check for *new* risks is a separate consideration altogether. The idea is usually that when you check your list weekly you can, at the same time, consider whether you have become aware of any new risks that you need to track. This is a reasonable approach, but it can be hard to find the time on a busy Monday morning to properly consider new risks (which are probably the ones that are hardest to spot – otherwise you'd have spotted them already). I'd recommend supplementing this with additional riskstorms, which might duplicate some of the ground covered by the original risk storm but will provide your best chance of identifying new risks that you should be considering.

9. The aim is to turn risks into actions – so track those too.
At the end of the day, anything that you change on your project will be done via an action. Even a regularly checked and well-maintained risk list will not achieve anything unless it results in you taking action and changing what you do.

The primary goal of a risk log is to turn risks into actions – transition indicator-checking actions to determine whether risks have been actualised, mitigation actions to help prevent risks from materialising, and contingency planning actions to prepare yourself for if they do. Seen this way, it's key that you integrate your risk log and action plan ('to do' list) well.

If you have a good working process for getting through actions, you may simply need to track the fact that actions have been transferred to your 'to do' list. Or you may wish to cross off mitigation actions as they're taken. There are many different ways that this can be achieved, and the best way to do it will depend on the exact processes that you use elsewhere. I recommend giving it some thought.

One major failing of a risk log occurs when, for whatever reason, the actions that you should have taken don't happen. Actions can get lost between the risk log and your action list, or you can defer copying them over until you're less busy, or they can make it over and languish at the bottom of your 'to do' list behind all the other super-urgent things that you have to do. Remember that, if you don't take the actions that your risk list has told you that you need to take, your risk list will be pretty toothless.

10. Work out how to make yourself check the risk log
The first thing that I look for when reviewing a risk log is whether there are any risks that have passed a transition date and not been updated. Risks that are weeks out of date are the most obvious sign that someone is not on top of their risk process.

If you have a risk log, an important pre-requisite is that you check it with the right frequency and act accordingly. You can have the best plan and format, but if you're not making regular updates it will all be for nothing.

You may be able to simply tell yourself to check your log, or you may have an infallible reminder system to which you can add an entry. For most busy project managers who are starting a project, though, it's not that simple. It's even possible to open up a risk log at the appointed regular time without focussing enough to make sure everything is correctly updated.

It's important in this case to have ways of incentivising yourself. Finding ways to convince yourself to do things is an important skill and I could write a whole other book on the topic, but for now I'll just say that I find having a useful, working risk log to be the best incentive to keep checking. So perhaps if you can kick start it, it will be easier to carry on.

11. Take it out!

'Less is more' has become something of a buzz-phrase. People are starting to understand minimalist design and the concept of a search cost. With a risk log, less is definitely more. If you have a very long table with umpteen risks that you don't need to worry about, you'll miss the critical ones. If you get used to scanning down a list of items that don't change, you'll be less ready for it when something does.

We've already looked at some of the types of risk that you can leave out completely. But it's not just risks that you can leave out. Lots of standard risk log formats contain a large number of columns – and books like this one will suggest you add even more! The more information that you have in your risk log, the harder it becomes to see the pertinent facts.

Make sure you remove not just risks that don't add anything but also unnecessary columns. There is often an 'owner' column – is this always you? Do the likelihood and severity columns change anything about how you manage the list? Would it be better to remove the risks that have low likelihood and severity and not use those columns? Do you have a RAG (Red, Amber, Green traffic light) column? Could you just colour the rows instead?

The other advantage of removing extraneous information is that your risk log takes less time to read and maintain, meaning that you have more time to spend properly considering the risks. Don't underestimate this as a benefit!

This isn't meant to be a prescriptive banning of all columns that I don't happen to like. It's *your* risk log; you need it to work for you.

12. Be specific

Fuzzy thinking is the bane of management. You have an inkling that something is wrong, or could go wrong, or just needs some attention – but there are pressing, concrete things that need to be done now, so you ignore your inkling and carry on. This is exactly what a risk log is there to counter. But if you go a step further and *write down* a non-specific risk, you won't be able to take the mitigation actions that need to be taken.

Each risk in your list must refer to a specific outcome that may or may not materialise. The more specific your risk is, the better your counter-measures will be. Favour 'Risk: specification will not be ready by 20th June' over 'Risk: specification will not be ready on time', 'Mitigation: run through deployment plan specified in Deployment. docx by the end of sprint 2' over 'Mitigation: run through deployment in advance of go live', and 'Contingency: ensure duplicate backup server (see Serverspec.docx for specification) ready in case of failure' over 'Contingency: backup server?'.

Your risk log is there to make life easy for you. It's there to do the heavy lifting of remembering what to do and when. If you leave ambiguity, you'll need to do more thinking at the critical points when you should be running through your list and sorting things out. Bring the thinking forward. Be specific.

13. When a risk is completely mitigated, you can remove it from the list.

The usual process for maintaining a risk log is as follows.

- Add risk to list.

- Fill in mitigation and contingency columns.

- Wait for risk to happen.

- Either execute any contingency steps or remove from the list.

In this scenario, the risk log is serving as a reminder that, if X (i.e. the risk) materialises, you need to do Y (i.e. your contingency steps). This is a useful function in this case – but not in every case.

For some risks, you will be lucky enough to be able to plan your contingency so that you have project success either way. If the risk is that your server will fail, and you set up an identical backup, you may have a risk as follows.

Risk	Transition indicator	Mitigation	Contingency
Server fails	Set up monitoring to check server is up	Service regularly	Set up identical server as backup – switch over if risk materialises.

Each of these columns translates into an action. You need to take a recurring action to service your server, an action to set up automatic monitoring of your server, and an action to set up a backup. Once you have taken these actions, your risk is no longer useful. You could keep a row to say 'If my server fails, switch over to the backup'. But are you going to forget to do this? No. Keeping risks that have been completely nullified in your list will just get in the way of risks that you do need to be monitoring manually.

If you use your risk log well, this will apply to a large number of risks. It may seem slightly pointless to set up a list, add risks, and then take them off again, but once again you have to remember that it's not having a list that is the useful output here – it's the process of risk management. If having a list has caused you to think of a risk and take actions to completely disarm it, it has been a great success, even if the list itself no longer exists.

Use ordering

I mentioned earlier that, if you need to RAG (Red, Amber, Green) your risks, you may wish to do so by colouring your rows rather than by adding another column. This gives you extra information without adding extra text that will take you longer to read every time you check your list (and if you're using your list well, you'll be reading it a lot, so this cost may not be insignificant.)

Another way to add information without text is through the ordering of your rows. This will draw your eye to the risks that you have decided need most attention. If you're using RAG, you might want the red risks first. Or you might prefer to have a sequential list of risks ordered by transition indicator date, so that you know which ones you need to worry about first. Experiment. See what works for you.

Share your risks with the business

Openness has its pros and cons. Evangelists believe that everyone should be open about everything, whereas pragmatists sometimes prefer to communicate only certain messages and avoid the difficulties that can be caused by exposing everything. Risks logs are a perfect example of this. Will seeing the risks that you are facing alarm your customers and cause them to insist that you prioritise in a non-optimal way? Or will seeing a risk list reassure them that you're on top of the project? Will it remind them that some risks are within their control and need to be managed from their side too?

In general, most businesses favour a shared risk list. It ensures that everyone is on the same page, that key information is not kept from people who need to know it, and most of all that nasty surprises don't occur. If you are trying to conceal the risk that the software might not be ready on time, hiding it will work in your favour if the risk doesn't materialise – but you'll be much worse off if it does. I believe it's almost always better to be completely frank. If you're unhappy with showing your risk log to the business, ask yourself why this is. Have you correctly identified the risks that face the project? Are you managing these as well as you can? Are you hoping to get away with being lazy and not taking some actions that you obviously should be taking? Even if you do decide not to make your list accessible to the customer, thinking like the customer might well increase the rigour with which you manage your risks – which should help ensure a better project outcome.

A word of warning, though. If you do choose to share, make sure you don't end up in a half-way house where, because you share your risk list, you omit risks from it because you don't want the business to see them. Having half a risk list is worse than having no risk list at all. In this case, consider very carefully what you are trying to achieve and how it can best be achieved.

When working hard isn't the answer

Hard work is generally considered to be a universal good. People who work hard have more success. Putting effort and energy into getting things right can only make them better. Staying a couple of extra hours to finish off your task will always result in more code (or at least better code).

Is it possible that working hard can ever be the wrong decision? I'm not talking about burning yourself out by working too much (although I would always recommend thinking hard about your own work-life balance). Instead I'm talking about taking a decision to work hard rather than plan your way out of a problem. If you have a deadline that you've signed up to and unexpected problems mean that it looks like you might not be able to hit it, dedicated types tend to get their head down and work hard. If someone asks for a challenging delivery, developers will often say 'yes' and pour out some coffee.

I do truly believe that this is an admirable reflex. The only problem is that it does not universally resolve your problems.

It can often seem that deadlines mean that there is no other choice. You are in a shot-to-nothing situation. If you work hard you might deliver what is needed by the business; otherwise you fail, and all bets are off. But this approach misses the fact that life goes on. There will be some consequences of missing a deadline, but civilisation won't collapse. If you work as hard as possible and you still don't make the release, something will be done and the business will continue as before. The world will not end.

The correct approach if deadlines are tight is to take the opposite path to hard work and stop work completely. Yep, you read that right. If you realise that life will go on if you miss your commitments, you can start to think about exactly *how* it will go on. If you plough ahead now and miss the deadline, what decisions will you have to take? Will you push back a release? Will you release an old function without so many features? Will you close down your business and start again? That last scenario is hardly ever the actual outcome, but what if it is? If you own up to that now and accept that it might happen, you can take steps to secure a better outcome.

Let's imagine that you must have a full functional release by Friday. The end users are expecting it, it has been wildly promoted, and sales will suffer if it is not there.

Now imagine that the development is not going to fit. You can add extra members to you team, you can work all weekend, but you'd still need a miracle to get it done. Hoping for a miracle is not good planning.

What options will you have if you get to Sunday night and the code still isn't done? There are a few standard options.

1. Email the users and tell them that you will delay the release.

2. Release a system with less functionality.

3. Don't email the users, but release a couple of days late and hope that they don't notice. Depending on the nature of your release it's very likely that they're not all sitting there waiting for the exact moment the new version comes out.

4. Release all the features but with obvious bugs that you didn't have time to fix.

5. Invent some maintenance on Monday and Tuesday to give yourself time to finish the code.

Obviously some of these ideas are better than others! And some are clearly not choices that I would advocate taking. The point I'm trying to make is that, by stopping and planning ahead, you increase the amount of choice that you have available.

For each of the options above:

1. If you're going to have to delay the release, users waiting for a key piece of functionality are going to be much more inconvenienced if they receive an email on the day that they were expecting to get on with working on the new system than if they've been given plenty of notice. By realising ahead of time that you're going to have to take this option, you can give them more choice and allow them to plan around the delayed release.

2. If you know you're going to cut some function, planning ahead will allow you to choose what it will be. If you blindly work ahead right up to your deadline, assuming that you will either make it or die trying, the missing functionality will be whatever stories didn't fit. If you start with what you have and work out what would be best to add, you can plan for a much more usable cut-down release.

3. If you know that you might be releasing late, you can work out what the consequences are for the people who needed the release on Monday. Should you email them after all? They will now have plenty of notice to work around the missing features. Could you put a message up on the system homepage so that they at least know what's happening when they go there? Adding in this message could be a better use of time than coding another story point, simply in terms of time saved in support calls.

4. This one's easy – by planning ahead, you can choose whether to have fewer well-tested features or a delayed release rather than a release riddled with bugs that will damage your professional reputation.

5. I hope no-one's seriously thinking of using this option! However, if you know your release will be delayed and there does happen to be some maintenance needed, you might be able to combine the two if you plan ahead. Leaving it to the last minute reduces your available options.

As with a lot of management problems, the challenge here isn't an intellectual one. In the cold light of day, most people can see that they should be stopping and planning rather than panicking and working on. It's harder when you're in the thick of it. As the leader, you need to make sure that your environment encourages people to stop and think. You need to make sure that you know when deadlines are tight (see *Monitoring* p100) so that you can help people to stop and think. You need to make sure you reward careful planning rather than lucky, risky, work-hard practices.

Explain the commercial reality

In organisations, lots of costs are hidden. It can be difficult to link what you are spending with what you need to achieve. It can be easier for people to find personal perks that they wouldn't pay for themselves than to invest the same amount of time making more money for the organisation so that they can earn a bigger bonus. This situation is lose-lose.

In a smaller company, you are probably best served by some form of "open kimono" accounting, tied into a profit driven bonus. You need to make sure it is clear to people that it is worth their while to do the things that make the company money. At a larger organisation you may find this more difficult to achieve. However, by being honest about what your constraints are as a department you can get your team bought in to what you want to achieve.

This is a separate idea to the one of articulating a vision for your team. A vision makes the best motivator and establishes what everyone should be trying to achieve. However commercial reality has to inform how to go about achieving the aims of the organisation. If one approach achieves the end slightly better but costs twice as much, it needs careful consideration before adoption.

Is burnout a management problem?

I've described how you should set goals and let people choose their own path towards hitting those goals. But at what points should you intervene? If someone is achieving their goals at the expense of someone else, it's a clear-cut case of a situation where it's your job to step in. There will be many other, similar cases – but what's more interesting are the times when things are less clear-cut.

If you're leaving someone alone to achieve their goals, it implies a certain level of trust that they will go about these in the way that they believe is best. If they choose to work early or late, at home or in between games of pool, you're giving them the freedom to experiment and do it their own way.

Looking at it this way, it seems that, if an individual chooses to overwork in order to hit their goals, then this is their choice. Most company reward systems means that working all hours of the day will earn you more money. If you go all-out to complete a project, you'll be rewarded with both positive feedback and more cash. So should an individual be allowed to choose this path?

I can understand the argument that says yes, they should be free to choose. Personally, however, I think that there's a very good reason to cut off this particular choice. If your incentives are set out to reward people who work harder, it's fair for people working to those incentives to assume that all-out, non-stop hard work is what you would most like from them. But as a manager, that's not what I want. I want productive, happy individuals who are hitting their short-term productivity goals and long-term career goals by doing a decent week's work.

Some people will enjoy their job more if they work a few extra hours to hit their goals. No problem! In fact, I applaud this attitude. Some people will achieve their career goals sooner by taking an interest in technology and reading around the subject at the weekend. Again, I see this as a positive step that will make them happier as well as more productive.

But routinely working 14-hour days and weekends hardly ever makes people happier. And it doesn't always make them more productive. It's not what I personally want for my colleagues. Telling people that you'd rather they didn't do something while paying them more for doing it is not always a very effective message. So I intervene quite strongly if I feel that people are overworking. I'm not prepared to remove personal choice (indeed, there's not much I can physically do to prevent people working till 2am), but I can make it clear what we as an organisation consider to be a reasonable effort, and what we consider to be beyond the call of duty. I can explain that I prefer working smart to working hard (I genuinely do!), and that it's almost always more important to stop and plan than to work blindly through the night.

It sounds like a contradiction, but I believe in trying to curb people's choice slightly in such cases. The unpalatable alternative is that they think you'd like them to be taking a certain course of action, when in fact that isn't how you'll be measuring them.

Developing your people

A common management objective is to 'get the right people on the bus'. Once you have them on the bus, however, you need to be thinking about how to make them better people – before they decide to get off again.

Providing career paths and upskilling is sometimes seen as optional. It's not uncommon to hear employers say: 'We can offer a great working environment – but limited development opportunities.' This will work if you accept high staff turnover. In general, however, high turnover leads to a weaker team, as the stronger and more ambitious team members are incentivised to leave and those who would be less able to get another position are the ones who will stay.

Developing people should always be one of your top objectives.

At some point in your career, you'll move from training your direct reports to coaching and mentoring them. Training and coaching are complementary skillsets with slightly different angles. Developing your coaching skills can be accelerated by taking on mentoring of someone outside your organisation.

This chapter covers the following ideas.

- **Make positive feedback easy**
  How do you get people giving positive feedback in the golden 5:1 ratio?

- **Align assessment of performance**
  Does your team's opinion of how they are doing match yours?

- **How to approach training**
  Training costs time and money. Find out how to get the right training at the right cost.

- **Embrace failure**
  It is only by failing that we learn.

- **Rules for accepting feedback**
  Ensuring that, if someone gives you some valuable feedback, they continue to do so.

- **Stretch goals**
  Just having goals is *so* last century.

- **Become a coach**
  A sports metaphor that's stuck: make sure you're training your senior team the right way.

- **How to both learn and deliver**
  You can have it all! Train new people while not letting delivery standards slide.

Make positive feedback easy

Guests from the chattering professions who visit the development offices are often astounded by how quiet they are. Techies are in general not a voluble bunch. Sure, get them down the pub talking about the Planck length or solving a puzzle and they can chatter away with the best of them. But software development work requires concentration. While chat can help spark ideas in other jobs, it's more likely to derail your train of thought if you're a techie.

When you're not chatting, it's hard to get into a culture of positive feedback. Positive feedback is one of the most motivating tools in any manager's handbook. As we've seen, some research suggests that you should give as much as five pieces of positive feedback for every constructive suggestion. While you as a manager may be doing your damnedest to instil a positive feedback culture, it won't get far if it's just you being positive on your own. Rather than generating a new culture, you're likely to be dismissed as a crackpot following the management rulebook.

The most innovative solution that I have seen for this is Softwire's custom tool for encouraging positive feedback. Known as Toast (since it allows you to 'toast' your colleagues and uses pop-ups that are sometimes also called 'toasts'), this little tool runs on your desktop above your other open windows. You can enter another employee's name and a description of what you'd like to praise them for. When you click 'submit', an email containing the praise is sent to the employee so they know how pleased you are with them. So far, so what? This isn't much different from simply sending a congratulatory email. The clever part is that *everyone in the company* gets a popup at the top of their screen the moment that someone is praised, making positive feedback really visible to others. Recently, a team racing against a hard deadline (in both senses of the word: challenging *and* unmovable) kept team motivation up by praising the hard work that each member was putting in, thus encouraging everyone else to also stay focussed.

Align assessment of performance

Most companies have an appraisal process. These tend to employ a top-down assessment mechanism, designed to feed opinions down the chain and feed into salary reviews. This is a good start. Done well, a top down assessment can help employees to understand their current place in the company and what they need to work on to progress to where they are going.

A better objective, however, is align understanding between the subject and 'the company', by which I mean the subject's peers, managers and managees. If aligning understanding is your objective, then allowing the subject to write the first draft of their own appraisal can often achieve this objective better than getting their manager to write it for them.

For self-aware individuals, not much editing is needed to their first draft to bring it into line with an appraisal that their manager is also happy with. These are the individuals who are able to sit down, think about what they do well and what they struggle with, and put together a review on their own. The process works excellently for these peeps, because it forces them to sit down and think about the longer term rather than the day-to-day. Then their views are reaffirmed by their manager.

Sometimes, more alignment is required. But this also works really well for the individuals involved. As well as a useful appraisal, they also get some training in how to think dispassionately about their abilities and put together plans to either build on their strengths or work to improve in weaker areas.

It can be easy to think that a top down approach achieves this too. 'We tell everyone how they're doing, and then everyone's understanding is aligned.' Sure, this approach works when you were already aligned – how could it not? But if your view *aren't* aligned – if you think an employee is close to failing because they're dropping loads of actions, and they think they're doing brilliantly apart from some unimportant admin – you simply telling them that is unlikely to change things. Chances are, you've told them that already. The message that they take away time and time again is that you don't respect the great job that they are doing and that you're just fussing over something that doesn't matter.

Only if you start by teaching them how to accurately assess their own strengths and weaknesses can you achieve alignment and hence improvement.

How to approach training

When I left university, I had no commercial programming experience. I'd spent two weeks at the tail-end of a summer job working on an online SQL training course, but I hadn't had a chance to put what I'd learned into practice.

I hadn't thought too much about what was going to happen at my new job, but there was one thing that I definitely expected. They were going to turn me into a programmer. I didn't know how to program, and they did. So there would be some training courses and practice, and suddenly I would know what to do.

Things didn't work out quite like that. Aside from some very basic training in the languages of the day (back then, ASP) I was put straight on to a real, commercial project.

This is how I learnt that working on a real project is better training than any training course. There's a reason why recruiters ask for a certain number of years' commercial experience. It's because you learn different things working on a project than you do from slogging through a course book. Working on a project is not a case of simply running through some example code by rote because the previous paragraph has explained exactly what it does. Real coding is messy, and it's learning to deal with that messiness that gives you the ability to do your job.

If you're commercially minded you will probably already have spotted that, what's more, if someone's working on a real project, albeit more slowly than a fully trained developer, you're also getting some real output.

There are a few provisos that you need to be aware of if you're thinking of training your team in a similar way. The first is that the real training and learning comes through a peer-review system where every piece of code written is reviewed by an experienced developer.

Many people think that peer review is useful for quality control. This isn't its main advantage. Thorough testing and QA is best for quality control. The *real* advantage of peer review is training. When you learn through peer review, you essentially learn by getting it wrong the first time. That's how we all learn fastest. But for some people, this can be an emotionally difficult thing to come to terms with. It's tempting to rail against the feedback to show that you *weren't* wrong. And it can be tempting for reviewers to take the easy route with those people, and pull their punches.

So the first thing that needs to be taught is how to accept feedback – how to think about what you've done and what you could do better as an intellectual exercise rather than a personal assault. Structure your appraisal process to reinforce this message.

Second, you need to make sure that you don't need to rework everything once it's been developed. That's obviously not a great use of time, and it's also soul-destroying for the person who's just invested all their energy in creating a sub-optimal solution. You can get round this by having several iterations of review for trainees, with at least one extra review at the design stage.

Embrace failure

Whatever we do, whether it's software development, building cars or policing a city, we are all trying to do it well. We are all trying to succeed. We set our sights on the end point, and aim towards it. We do everything in our power to avoid failure. We carefully manage a list of risks so we can deal with the unexpected and stop it from throwing us off course.

Some organisations go further. You'll have heard the expression 'failure is not an option'. This seems like a reasonable way to go about business, surely?

It does – until you examine the concept more closely. What would you do on a project if failure *really* wasn't an option? First, you'd make sure you got the easiest project you could. Let someone else take the risk of failing. Then you'd try and find a role that you'd done before and stick with that. If you were a developer, you'd use the coding language that you've always used. You'd never question the company standards that have delivered successful projects in the past. In short, you would never try anything new. Worst of all, you, and everyone on your team, would try to tidy up after a project to make it look like the best project that you'd ever run. You'd hide the things that could have gone better. You'd ignore the niggles that made you slower.

Is that so bad? You do it well, everyone feels good about it. What's the problem? The problem, which may at first sound somewhat tangential, is that the world doesn't stand still. You are always competing with someone else. If you're a software development company, you have competitors. If you're an in-house team, you might lose your jobs to people who have new and better ways of doing things. You might start to lose work to external companies who can deliver more efficiently. New companies will start, new ideas will develop, and new ways of working will change the competitive space.

Planning to avoid failure is the antithesis of learning. If 'failure is not an option', you need to reduce risk. New things are risky. Furthermore, improvement comes from looking back at what could have been done better. We learn from our mistakes. If we churn out perfect, identical project after perfect, identical project, we learn nothing. But if we can identify where we have failed, we can deliver a better project next time.

In order to truly embrace failure, you need to detach it from blame. Often in management, the trick is to do the opposite of what comes naturally. If someone has screwed up on your project, of course your instinct is to blame them. If someone points out that you weren't so hot yourself, of course you get prickly and defensive. It takes work to overcome these reactions.

In order to lead a team, you need to start by modelling the behaviour that you want to see. Be open about the things that you personally could have done better. Encourage people to tell you when they spot something that you didn't do so well. Say 'thank you for the feedback' (see *Rules for accepting feedback* next). As you practise being open to the fact that you could have done better, it gets easier and easier. As people see you reacting in this way, they'll find it easier to try it themselves.

Rules for accepting feedback

As you may have gathered, I'm a great fan of continuous improvement. Feedback is the lifeblood of continuous improvement. Whenever you undertake something, you should be thinking carefully about what could have been done better; this will provide you with some great improvements without any outside help. But there will be a whole class of things that you could be doing better that you won't be able to spot. To a certain extent this is self-evident – if you knew you could do it better, you would already be doing so. Some things that you could do better will be that way only because other people have reacted in a certain way to something that you did – they found your manner abrupt, or they didn't understand what you were asking. In order to become the best manager that you can be you need to find out this information, and act on it.

To get to the position that you're in now, you will have already had to learn how to act on feedback. You'll have picked up both specific skills and ways of working and behaving. At the bottom of an organisation, feedback is usually relatively easy to come by: if you do something wrong, you'll find out about it; if you upset your boss, they'll tell you. You'll also be keen to ask for feedback, and your manager will be happy to give you it.

Things change a little when you are the one in charge. It's no longer obvious to others that you want people to feedback to you frequently and honestly, even if you state as much in departmental presentations. It can also be easy when you are rushed and busy to respond to feedback curtly or peremptorily, even if you do find it useful and later go on to act on it. Or you may have an emotional reaction to feedback, especially if you do feel deep-down that you have done something badly. Add that to the fact that giving useful, honest feedback is actually really hard, and you might find that people fear giving you honest feedback in case they upset you.

I once blogged the following rules for accepting feedback.

1. Whatever the feedback is, immediately say 'thank you for the feedback'. This shows them that you appreciate their taking the time to help you, and will mean you get more feedback in the future.

2. Before disagreeing (or agreeing!) with the feedback, take 15 minutes, or however long you need, to absorb the information… or calm down.

3. Only then think about whether you agree with the feedback or not, and what you plan to do about it.

4. Feed back to the feedback giver on how useful their feedback was! Remember, this is something that they probably didn't find easy, so take the time to let them know how they did and provide any constructive comments you have to help them get better.

It's not necessarily the case that all feedback you receive will be equally useful – if you let people know which bits were most helpful and which were less so, it will help them to practise continuous improvement on their feedback giving.

Stretch goals

As I mentioned in the goals section (p94), setting goals is difficult. Make them too easy, and good developers achieve the goals almost by accident; even worse, bad developers have no pressure applied and may end up missing even the easy goals because they realised they could slack off. On the other hand, goals that are too hard are even worse. If they are acknowledged as impossible they provide no motivation at all, and if a team works flat-out to hit a goal and still doesn't achieve it, they will go home dispirited despite having done an amazing job.

As with most things, the key to setting good goals is to think hard and carefully about what you want and are able to achieve, and write it down accordingly. But I have another tactic that I like to use to take the pressure off goal-setting. I set two groups of goals – bedrock goals that must be hit for the week to be considered a success, and 'stretch' goals that indicate performance above and beyond the bedrock goals. This removes the pressure of having goals that are so challenging that you might fail, while still giving you something more to aim for.

I actually use a similar tactic with my (very busy) 'to do' list. I read somewhere that you should only ever have three important things to do in a day. This sounded like madness to me when I first read it, and I still do have a lot more than three things that I need to achieve each and every day. But if I ignore the things that will just happen – such as meetings that I've booked in or emails that I'll read – I find that identifying a smaller number of things that I absolutely *must* get done (sometimes it can be a little higher than three...) allows me to make sure that I don't miss those out for something less important.

This allows me to feel that my day has been a success, even if I didn't complete everything on my list.

Become a coach

There was a time when you were expected to turn up, work hard and do your job – and do that same job for the rest of your life until you retired at 65. Nowadays, everyone starts out expecting to do better year after year. Many people start out with the ambition to get to the top, and anything less seems like failure. To most people, work is a progression, not just a day-job.

This isn't a bad thing. All jobs have tedious parts to them, and repetition makes even the best vocations boring, so having something to aim for and achieve in your career, especially if it's tailored to you as an individual, makes working life more fulfilling.

But seeing your career as a training game means that you need something more from your manager than just instruction. You need them to be a coach.

I'm sure you can remember someone seminal in your life. Almost no-one can get to where they want to be without a little help. It may have been your manager, or it may have been another colleague or even someone from elsewhere in your life – a partner, or a friend. In fact, someone who isn't directly involved in your work can sometimes be more helpful, as they don't have any preconceived ideas about what you should be doing or where you should be going. Your manager might think that you want to be just like them, but without the pitfalls that they think held them up along the way.

Now that you're higher up in the organisation, you need to give this back to everyone else in your team. Your first job is to be the mentor for the key people in your organisation. It can be a good idea to be the mentor for some of the younger rising stars, too – it's good practice for you and will be very beneficial for them. Once you establish this culture, you'll find that a lot of your core team will want to mentor too. Then you can make it clear that you consider this an important way to give back.

Mentoring is one of those activities that you learn best from watching an expert doing it, and then, through years of experience, figuring out what works best. There are some good books on the subject out there, and if you read closely and take notes these can give you a useful head start.

Here are few good tips for starting out as a mentor.

1. The mentee needs to be ready. You cannot mentor someone who is not interested. This activity has to be opt-in only, otherwise you are wasting your time.

2. It needs to be clear that the mentee must lead the relationship. They must arrange the meetings, set the agenda, and take notes. The mentee must be the one who decides what they want to get from the relationship and how they would like the mentor to help them.

3. If you and the mentee don't work closely together, spend some time getting to know each other first. Find out about what makes them tick – what do they enjoy doing?

4. After getting acquainted, setting goals should be one of the first activities you undertake. Without goals, you will find it hard to make concrete progress; your mentoring chats may be pleasant, but you'll find that they swiftly descend into little more than chats.

5. As a mentor, you will usually be the more experienced party. Try not to judge. If someone doesn't want to take the same path as you or wants to do something you consider bizarre, that's their

decision. Answer their questions, offer suggestions, but don't ever tell them what to do.

6. Mentoring can sometimes stray into personal topics. As individuals, we are all shaped by our past and our personal lives. Mentor and mentee may well have different ideas about what's acceptable, so agree upfront what topics will and won't make suitable mentoring material.

7. You need to establish a personal relationship in order to mentor effectively. Try to have your meetings somewhere away from your usual workplace – maybe in a local coffee shop, or at a bar after work.

8. At some point, the mentoring relationship will end – maybe after a set of goals have been achieved, or if you both find that the relationship is not delivering what you were after. Don't be tempted to try and extend a mentoring relationship beyond the end of its natural life.

Training works best when it plays to the natural ways in which people learn. Mentoring and learning on the job can often deliver faster benefits than training courses or book learning, and can be very enjoyable to boot.

How to both learn and deliver

Being able to learn is a key part of being happy. It is also widely cited as a good reason to take a job – books such as the cheesily named *The Start-up Of You* by LinkedIn founder Reid Hoffman advise you to consider how much you will learn as a key criterion in deciding whether or not to take a job. My favourite maxim (probably because it involves a rhyme, as all good maxims do) is from *Never Eat Alone* by Keith Ferrazzi: 'Learn in your twenties, earn in your thirties'.

On the other hand, it's no longer acceptable to turn up to a job and expect to be trained. Competition in the technical job market is fierce, and even if you're applying for a job that doesn't require experience you may well lose out to someone who has been learning in their own time and has examples of websites or applications that they've produced on their own initiative.

There's also the question of whether it's the responsibility of your employer to train you. There's no doubt that as an employer it's useful to have people who continue to learn – but it is arguably *more* useful to the employee themselves. This is where we see a split between technical roles and other, more traditional jobs. Technical guidance manuals such as *The Clean Coder* by Robert C. Martin are very clear that training is not your employer's responsibility. Martin suggests that, in addition to your 40 hours' delivering for your employer, you should spend another 20 on personal development and training through effortful practice.

If a developer wants to get on, the more they take an interest in the subject outside their day job – getting involved with side projects, meetups, forums and discussions, or reading relevant books – the better they will do. As a leader, you should be clear with your team about this. Otherwise it can seem to those with the wrong mindset that some people learn and improve almost by magic.

Of course, there's also competition among employers to attract and retain the best developers. As an employer, you would probably like it if you could employ a team of self-driven coders who do 20 hours' training in their spare time – but in reality you are going to need to provide some training too.

A good balance is to ensure that people are always working just outside of their comfort zones. If you make sure that developers are being stretched by new technologies, domains, application types and roles, they can learn and deliver. It's a fine balance. Being stretched just the right amount can result in improved delivery, as focus is required and boredom doesn't have a chance to set in – but get too far out of your competency, and delivery is going to suffer.

If you're the person in charge, constant stretching and learning also makes your job more difficult. The landscape is always changing. You can't let your team get on and deliver in the same way that they have always done; you need to keep re-planning and adjusting. You also need to be on the lookout for anyone who has been stretched too far. Don't forget that keen, ambitious people are often afraid to admit when they are in over their heads. Look for signs of overwork and don't be afraid to jump in and give extra support.

Empowerment

You haven't got to the position that you're in today without having learnt how to delegate. Empowerment is the close cousin of delegation. It's the bottom-up complement to delegation's top-down approach.

Empowerment rests on the basic premise that 'responsibility is taken, not given'. You'll know which of your reports will take responsibility outside of their official role, so you don't need to be the one picking up the pieces. Conversely, even people who are formally given responsibility can refuse to grasp it and fail to deliver what's required of them.

Encouraging empowerment is a tricky business. It's almost an oxymoron. One of my favourite jokes (in fact, it's a true story) is that once, in a senior management meeting, someone in all seriousness asked: 'How can we make a grass-roots movement happen to improve quality?' If it comes from the top, it isn't empowerment.

Like all of us, I am learning every day – and this is an area in which I feel I have lots more ground to cover. In this chapter I'll share what I have learned so far.

- **Talk to the duck**
  Q: When is a manager like a rubber duck?

- **Don't talk too much**
  Call it 'active listening', or 'using silence', but the less you talk, the more your managees will.

- **Explain the trade-off between micro-management and taking responsibility**
  If I can be sure you can deliver, I can leave you alone to get on with it.

- **Top-down vs bottom-up**
  How do you get people to reach up, rather than you having to pass down?

- **Let your managees lead meetings**
  I'm sure you're used to being the one in charge, but if you want your reports to take responsibility, you need to let them lead.

Talk to the duck

'Rubber duck' technique is one of the innovations that I have seen introduced by a project manager who was willing to try something new (see *Trying something new* p175). He didn't invent the practice, but no-one works in a vacuum; the best project managers build up their skillsets through others' experiences as well as their own.

You know that feeling you get when you go to ask someone a question, and, just by framing the question in your head, you find you've worked out what the answer is? Lots of managers have to deal with those sorts of questions day in and day out, being interrupted in their busy schedules just to find that they're not actually needed.

Introduce the duck. He's just like any other manager, but he doesn't get grumpy if you go and talk to him on his lunch break, and he doesn't interrupt you just when you're getting to the 'oh hold on, I know...' moment. In fact, he waits expectantly for it.

The duck is the perfect manager that lets you work out the answer yourself.

Following on from the successful adoption of the 'rubber duck' technique, I've started training project managers with a new device that I call the 'reverse duck' technique. Not only does it help them to frame their own questions and solve their own problems, but it also teaches them the underlying basics of project management. Reverse duck technique can also be practised using a duck instead of your manager. The difference this time is that, instead of asking your duck the question you want to ask your manager, you think 'If I were to go and explain to my manager what I'm doing right now, what questions would they ask me?'. Project managers quickly learn how to present arguments that don't need to be questioned by senior management – *and* they learn useful questions to ask members of their own team.

Don't talk too much

Following on from the 'rubber duck' and 'reverse duck' technique, there's a third way to use the concept of a silent rubber duck solving your team's problems. This is to act like one yourself. Classic management books such as *How To Win Friends And Influence People* advocate the power of silence and listening, rather than talking and telling.

With technical people, this skill is both more important and more difficult. In a one-to-one status chat, developers can be hard to bring out of their shell. And the less they say, the more tempting it is for you to fill the silence with helpful suggestions and comments. However, because you need your developers to be creative and pro-active – much more so than with less demanding jobs – it's vitally important that you get them making their own decisions and speaking up when they have new suggestions of ways to approach problems. If you don't encourage them to tell you, you run the risk of their implementing something that you didn't know about and will find out about too late – or, even worse, you'll miss the opportunity to utilise their talents and get a better solution. The more you suggest things that they can do, the more they'll feel like they've been given a set of arduous instructions, and the keener they'll be to get the hell out of your status chat.

Some of the most important information I've learned from my team has been elicited by simply shutting up (this is a management skill that I've had to do a lot of work on, as my natural inclination is to offer helpful advice at every juncture: in my status chats with one of my senior project managers, I have to add a reminder to the top of my weekly notes to let him talk first and to not talk over him). By getting together some new people-managers in a room and sitting by while they asked and solved each others' questions, we got into some of their deepest concerns and extracted new ideas that would never have surfaced if I'd just sat them all down for a lecture.

Explain the trade-off between micro-management and taking responsibility

I don't believe in micro-management. Lots of people think this will make working for me easier. In fact, it makes it harder.

If you are micro-managed, you may have more or less work to do but your job is essentially easy (or not very challenging, depending on how you look at these things) because you only need to work through a set of pre-defined tasks. Once these are finished, you have done your job. You may find it unpleasant to have to keep quiet while someone tells you to do the wrong thing, but, if you can cope with that, you need never exercise your brain again.

If your manager is steadfastly refusing to micro-manage you, your job gets a lot harder. You need to deliver the release by the end of the week. How do you make that happen? You need to make sure that there's a testing plan.

Not being micro-managed doesn't mean that your manager won't help. Helping is exactly what your manager is there for. What your manager is *not* going to do is take your decisions for you. The difference between 'Do X' and 'If you are asking my advice, I would do

X' is subtle but profound. The first is an instruction, given on the understanding that you will do it and you will both live by the consequences; the second admits that it might be wrong. An even better non-micro-management answer is 'X will achieve Y, A will achieve B – what achieves our main aims best?' – Or even 'Go and ask your duck' (see *Talk to the Duck* p155).

Taking responsibility is quite difficult when you're not used to it. It's even more difficult when you're not used to it and still in need of training. Our school system (certainly in the UK, and, I understand, in the US too) doesn't train you for this in any way. At no point are you instructed to go and work out how to write an interesting essay on a topic of your choice. Instead, you are given a reading list (the same as everyone else), told some facts, and hand-held through writing. But if you're going to be successful in your career, you can't wait for someone else to do things for you.

Top-down vs bottom-up

Traditional management is 'top-down'. Decisions are made at the top, and flow down throughout the organisation. Tributaries are added on the way in the form of mid-level managers. The flow is downwards.

But this style of management doesn't work for a technical organisation or department. Leaving aside the fact that the people at the top are not best-placed to make a lot of the key decisions, developers are not the type of people who passively sit around and take instructions. In fact, many developers would rather leave and work somewhere else than operate in that kind of environment.

Typical management techniques in a top-down organisation include shouting, punishment for misbehaviour (removal of perks, or being put on an undesirable project), and performance-related pay. These all have only a limited effect on technical people.

The best way to describe the kind of environment that you need in a technical set-up is the polar opposite of this – a 'bottom-up' approach.

In a bottom-up approach, ideas and inspiration travel upwards. Information flows up the tree from the bottom. The people at the top are merely collators of the information from lots of different roots. Looked at as an information problem, this makes a lot of sense (and techies like things that make sense). The people who are doing the job are the people who know best how it can be improved. There may be costs associated with this that are unacceptable to the business but, by making sure ideas flow upwards, costs from different teams can be weighed up against each other.

Motivational drivers in a bottom-up environment include enjoying the challenge of your day-to-day work, taking pride in what you accomplish, and being respected by your peers and managers. Developers thrive in a bottom-up environment.

Let your managees lead meetings

Gone are the days when companies were ruled with an iron fist. The old way of managing involved one person at the top making all the decisions, and everyone else implementing them, possibly via a series of middle-managers. As long as everyone did what they were told, everything worked well. Management issues were mostly to do with how you got people to do what you had decided that they needed to do. A particular type of aggressive, shouty management got the job done best.

Empowerment is a word that's bandied around a lot nowadays (see *Empowerment* p166). I think of it as the exact opposite of the 'aggressive and shouty' school of management. It's possible that the old-school style of management still does work in, say, a traditional factory environment, but the emergence of new production methodologies, such as Toyota's Lean, suggests that it no longer has a place even there.

The problem with one person making all the decisions is that, rather obviously, one person is only one person. The limit to how much knowledge one person can usefully act on may be quite large, but it is still a limit. Distributing the mental processing across everyone in the organisation rather than getting one individual to do it all just seems sensible.

And there's another reason to get more people involved in the thinking process: people think differently. There are even benefits to having two people, rather than one, think about the same thing; they might come up with very different ideas.

I see an organisation as a series of 'thinking parts' working together to achieve an overall aim. The people who are working on specific deliverables do the thinking about the detail of those deliverables – they are simply the best-placed people to do this! Other people will do their thinking at a level that agglomerates data across a group of deliverables; they can form general ideas and principles from this, and, because they have all the information, are best-placed to do so. But they are still not best-placed to decide on the detail, even if they are considered more 'senior'.

However, people within a company are still human. This means that they often stray down the easier path of simply doing what they're told and no more. It's an approach that's actually fairly well-accepted in our society (and even espoused by poor managers). So prevalent is this concept that people often do what they *think* they're about to be told to do, even if they don't feel threatened or coerced. It's just quite an easy way to do your job.

The best way to tackle this issue is to try and reverse the usual management relationship. Your manager may set goals for you (or help you to set your goals), but once you know what you're aiming for it's your job to work out how to get there. You may need help and advice from your manager, but not because they're telling you what to do – rather, because they're a resource that's available to you to use.

I have a meeting-plan format for two-person status chats that I use to help underline this idea. The key component is to get the more junior member of the chat to lead the meeting. This seems back-to-front for most people. They are used to a status chat being a question of them reporting upwards; when I ask them to set an agenda, they don't know what to say. But once someone is happy with this format, I'll get them to email their agenda in advance; I'll also ask them to demonstrate good action tracking afterwards by emailing through the list of what was agreed in the meeting. This gives them a starting point for the next chat.

This can be quite scary as a manager, because someone might not ask you the 'right' questions. But provided that the brief is clear, either someone is not going to hit their targets (in which case you can – and must! – intervene), or they will hit their goals despite doing something that you consider suboptimal. This is the point at which you need to let them make their own decisions; otherwise you'll undo all the good work of setting up an environment in which they *can* make their own decisions. 'You can make your own decisions except when they disagree with mine' is simply not effective.

Look to the future

So you've improved your management, leadership and people skills and empowered your team – time to sit back and have a cup of tea. Or is it?

'Continuous improvement' is a fashionable phrase in software management (possibly as an analogy to continuous build or continuous delivery). A healthy organisation is constantly getting better. No set of practices is perfect, and standing still will allow your flaws to fester and take over; at best, you'll be left behind as your competitors find and develop new techniques that allow them to be faster, more accurate and perhaps even more fun.

Another reason for looking ahead is to make sure that you keep growing your top people (these objectives are often linked). Remember: without your team, you are nothing!

- **Recruiting technical people**
  How do you work out who are the superstar coders?

- **Train technical leadership**
  Just being good at coding isn't enough for the leader of a technical team.

- **Make sure people try new things**
  'If you always do what you've always done, you'll always win what you've always won.'

- **Create a technical career path rather than just a management one**
  Don't lose the superstar coders you worked so hard to recruit!

- **Reward using things other than money**
  What do people really want?

Recruiting technical people

Recruiting technical people is one of the hardest things that you will have to do. The performance gap between the best and worst developers can be tenfold or higher, so smart recruitment decisions can save you a lot of money. As with any type of recruitment, the difficulty lies in developing an assessment system that will allow you to assess, in the space of just a few hours, whether developers will perform well over months or even years.

There are two main things that you need to recruit for: **Ability** and **Attitude**:

1. **Ability**: being able to deal with complex thinking and to come up with the right technical paradigms.

You'll hear a lot of people just referring to this skill as 'being smart', but you need to be more specific than that. This is a certain type of smart. Lots of development companies are happy to recruit people from all university disciplines. If you're not pre-selecting by degree course, then you need to test each applicant's ability to think like an engineer. Well-designed interview questions help to identify people who break a problem down into abstract components and construct a simple algorithm. There are lots of different types of intelligence. You aren't looking for people who can think around the problem and spot possible social consequences, or think about what would happen if you varied the conditions slightly.

In most jobs, a "general intelligence" best described as ability in mathematics and English language is what is preferred. This is a good approximation for a developer, but sometimes it's not enough. And sometimes an applicant can have the right type of intelligence without having the general type that would be valuable in another role.

2. **Attitude**: a passion for software development, the determination to stick with the task at hand, the ability to take feedback well (to name but a few key qualities).

Attitude is by far the harder of the two things to recruit for. Even simply defining what you're looking for is difficult. It encompasses the following facets (some are essential; others would be beneficial).

- Interest in software development: reading about industry news and new languages and techniques.

- Enough passion to 'play around' with technology in their spare time, essentially training themselves on their own initiative.

- No fear of failure. Technology is still developing, and those who can only learn from others rather than by trying new things for themselves will always be at a disadvantage.

- Stamina. Software development is difficult, and it can be hard to keep plugging away when you're having difficulties with a specific issue.

- Emotional resilience. You can only improve if you can enthusiastically embrace feedback and learn from what you didn't do so well. If your pride can't cope with not being an expert, you'll fail before you've even started (although you may still be able to remain in the industry for years and years, bemoaning the young upstarts who always seem to get promoted above you).

- A willingness to share learning with others. It's not much use to an organisation to have someone feverishly learning about new technology and then keeping it all to themselves.

- Being a good team player.

- Time-management skills. Some technical jobs don't require this, but in general a developer who can manage a workload of several different tasks at the same time will be able to get through a lot more code a lot more quickly.

- The ability to take responsibility. In a purely Agile team this is essential, and it's still desirable in a team with a more traditional project manager. Essentially you want developers who say 'I see a problem, let's go fix it' rather than 'I see a problem – let's hope someone else deals with it'.

And yet attitude is the more important of the two. A study cited by Shawn Achor in his book *The Happiness Advantage* identifies positive attitude as being responsible for 75 per cent of success (with intelligence making up the remaining 25 per cent). My own qualitative observations of developers who succeed and those who struggle is that it's passion and attitude that make the biggest difference. It's common to put differences down to intelligence, but this rests on backwards deduction from the fact that the programmer is struggling: 'Vicky isn't doing so well at picking up C#, maybe she's not as smart as we thought.'

Most articles on technical recruitment contain questions that will help you to assess a candidate's ability to output good code. The second criterion will also play into this: someone who's studied hard and spent their own time investigating better ways to do things – i.e. displayed the kind of attitude you're looking for – will be better equipped to complete the tasks you set them. It's very hard to determine directly whether someone's attitude shapes up. So let's start with the easy option, and take a look at some interview questions that can help you assess someone's ability to write good code.

Pseudo-code examples

Lots of technical organisations understand the value of hiring graduates. Often, this is the most cost-effective way to recruit; universities are very helpful in providing ways to get in touch with graduates (mailshots, recruitment boards, recruitment fairs, milk-round events, etc.) and you can choose to target specific universities, which helps to ensure that you get applications from 'generally smart' people.

The problem is that these graduates don't necessarily have any coding experience. Sometimes they have personal experience with languages that don't have a commercial application; university computer science courses are often several years out of date in the languages that they teach. So if you set a test in, say, Java, you might seriously advantage someone who has come straight from a job using Java over someone who has no commercial programming experience at all. In short, you might end up testing only *what* people know, rather than their ability to apply it.

A good way to get around this problem is to use tests designed in languages that no-one can have a head start with – ones that you've made up.

Here's an example. Say you can control traffic lights using the following commands.

```
Turn-on (red|green|amber)      Turn on the
specified light

Turn-off (red|green|amber)     Turn off the
specified light

Wait (time)                    Wait for the
specified amount of time

#[label]                        A label in the
program

Goto [label]                    Move to the
inputted label and continue executing from there
```

You can then write a simple program to implement a standard traffic light algorithm as follows.

```
#Start

Turn-on (red)

Wait (5 minutes)

Turn-on (amber)

Wait (3 seconds)

Turn-off (red)

Turn-off (amber)

Turn-on (green)

Wait (5 minutes)

Turn-off (green)

Turn-on (amber)

Wait (3 seconds)

Turn-off (amber)

Goto Start
```

Of course, you'd want to make an interview question much harder than this, but this example demonstrates the basic premise. When asking the questions, it's important to spend a lot of time talking through the candidate's answer to get a feel for how they are approaching the problem; this is much more useful than just leaving them alone and seeing if they can solve the puzzle by themselves. After all, in the workplace, coding is not a solo activity.

Asking these type of questions gives excellent results in terms of recruiting people who have the raw ability to pick up coding. But how successful they are beyond that is determined by their attitude and endeavour.

Simple coding questions

If you only recruit software developers who have experience of a particular technology, you have more options open to you. One approach that is increasingly gaining currency is the idea of asking very simple programming questions. You might think that, in order to find the superstar coders that we're all searching for, it would be more appropriate to ask a really hard question. But the problem with that is that the most difficult questions take a long time to answer – time that is probably not available to you in an interview. It's thought that up to 99.5 per cent of programmers can't answer simple questions well and quickly, so these questions can be more revealing that you might expect.

Some commonly used questions include the following.

- Reverse a string in place.

- Reverse a linked list.

- Binary search.

- Find the longest run in a string.

- Write a function that determines if a string starts with an upper-case letter A-Z.

- Write a program that prints the numbers from 1 to 100, but for multiples of three print 'Fizz' instead of the number and for the multiples of five print 'Buzz'; for numbers which are multiples of both three and five, print 'FizzBuzz'.

Joel Spolsky runs a software company called Fog Creek in the US. He blogs regularly about programming topics, often with a great deal of polemic. Spolsky is adamant that the only real test of a good programmer is whether they can understand C style pointers. As a developer who started out just as C and C++ were coming to the end of their heyday, I do have some sympathy for his position. I remember the level of incomprehension that it was still possible to fall into even after you understood, and could repeat, the definition of what a pointer was. Once you're familiar with the concept there's a kind of intellectual beauty involved in manipulating pointers; it gives you a feeling similar to the one you might get from working out complicated mathematics. The problem is that pointers are no longer widely used in modern languages; to me it feels wrong to interview someone based on a concept that they will never need to use.

Spolsky disagrees. As he argues, 'I want my ER doctor to understand anatomy, even if all she has to do is put the computerized defibrillator nodes on my chest and push the big red button, and I want programmers to know programming down to the CPU level, even if Ruby on Rails does read your mind and build a complete Web 2.0 social collaborative networking site for you with three clicks of the mouse.'

Discuss some code that the applicant has already written

Another option is to ask the candidate to critique a piece of code (possibly with some deliberate mistakes), or to have them bring in a piece of their own code for you to take a look at (combining these two techniques can help prove that it really is their code and not that of a friend).

A key question that isn't often asked is 'What other ways are there to code the same thing, and why not do it that way'? In order for a developer to fully understand that what they've written is correct, they need to have a good grasp of the other options that were available to them. Programming scenarios are very subtle; an approach that works in one scenario might, in another very similar scenario, not work at all.

Attitude questions

Recruiters everywhere also attempt to ask the questions that will get to the heart of someone's attitude to programming and to life; quite simply, we need to know whether they will fit well into our organisation.

My favourite example of such a question is 'Tell us what your friends would say about you if we asked them'. This is a great question. The candidate is forced to think about how they come across to others – and they're unlikely to have a prepared answer. In fact, it's unlikely that having a prepared answer would be helpful; this isn't a question where you can earn a box-tick just for using the phrase 'leadership skills'.

Another question in this vein is 'Tell me about a problem you had with a boss or teammate'. The question allows you to check that the person you're hiring has a good idea of how to go about resolving problems.

To specifically gauge someone's level of passion, and their ability to explain, asking them to talk about a previous project or to present on a topic of their choice can be illuminating. And if you're looking for programmers who spend time outside of work thinking about programming, why not ask them directly what books, blogs or internet sites they read.

Many employers put a lot of weight on these non-technical questions; if you do badly at one of them when being interviewed by Joel Spolsky, he will score you as 'NO HIRE'. But the problem with linking the outcome of an interview too strongly with 'attitude' questions is that good programmers can often be quiet and not inclined to shine in these situations. In addition, if you are interviewing young people without a great deal of work experience, it's possible to misjudge their potential based solely on their answers to questions. To put it another way, if these skills aren't present when you hire someone, they can be taught; especially if you're bought into training employees on a continual basis.

The only real way to assess how someone will behave and perform at work is to *watch* them at work. If your recruitment policy allows it, offering a trial, rather than a permanent position, can help you spend a few months establishing whether someone really will fit into your organisation.

Something that is not always mentioned in conjunction with recruitment is staff retention. This is a bit odd, because for everyone person you retain, there's one person less that you need to recruit. In fact it's even worse than that, because the people who have the get-up-and-go to leave for a better job are often the people you really want to retain. But I won't talk about how to retain your staff here, because most of the rest of this book is concerned with how to make a great environment that people are productive in – and that's also what makes a company a compelling place to work. People enjoy doing things when they are doing them well.

Train technical leadership

You need to ensure that there's a technical career path as well as a management one open to your manages. Many companies nowadays ensure that there's a path by which their technical people can reach the very top of the company, whether this is via a Technical Advisory Board (favoured by companies such as Thoughtworks and UBS) or by other means.

Creating the technical path and putting people on it isn't enough. You also need to train people to fill the relevant roles. A common misunderstanding of the technical lead role is to see it as a sort of glorified developer, who gets to tell the other developers what to do. While a progression to project manager is seen as a move to a different role, the role of the technical lead can be seen as 'more of the same'.

But in fact there are key differences between a developer role and a technical lead.

1. A technical lead needs to take responsibility for the technical integrity of a piece of software. Although there will be a project manager who is responsible for delivery to the business, they may well be either non-technical or too busy elsewhere to personally ensure the quality of the deliverables. So the buck for technical quality stops with the tech lead. A developer needs only to code their own components well.

2. The technical lead must know about all areas of the delivery, including testing and rollout strategy. Unless you stick with a single technology and are lucky enough to not have to take advantage of any new advances (which in my experience is hardly ever), the tech lead has to find time to learn on the job about new technology and how to implement it. A developer will often work on one particular area, usually one in which they have specific expertise.

3. The technical lead needs to take responsibility for the output of each developer. They cannot say 'But George coded that part'! If they're going to be the tech lead for a long period of time with the same developers, they'll also need to be able to train and develop them in order to get the best long-term output. A developer only needs to be responsible for their own code.

4. Conflicts may arise around the best way to do something. Indeed, in a healthy environment, debate will be nurtured and encouraged. The technical lead needs to not only be able to calm down any heated tempers and reach a resolution but also to facilitate the participation of everyone in the team, hear the views of those who don't shout as loud, and keep their own temper long enough to hear out any developers who are contributing in less constructive ways. An even-tempered developer can keep their head down and stay out of the way of any conflicts.

Luckily, it's easy to train for a technical lead role, especially while working as a developer.

The first area to worry about is self-leadership. It's easy to look to others for direction, but when you're working on a piece of code and there's a trade-off to be made, a training tech lead finds out the high-level parameters of the project and works out their own answer. You'll become a much more valuable member of the team if you're reporting in with solutions (for review, of course) rather than with problems.

1. Customer interaction can be a good way to step up and show your leadership (see *Be nice to the customer* p67). There will almost always be non-technical restrictions that you have to overcome in order to get to the right technical decision (of course, making sure you have a good relationship with your project manager can be a good way to get what you need approved without having to do the dirty work yourself).

Setting yourself up as an expert in a certain area can help you to train in enabling others and setting technical direction. It will also teach you how to ramp-up to the level of expert in a new technology in order to be able to take the right decisions for your team.

Taking responsibility is one of the easiest skills to practise – and you don't need anyone's permission to do it. It was one of my great 'd'oh!' moments when I realised that responsibility is taken, not given; you may not get credit for it, but making sure you get credit is a completely different skill (see p26), and one that you can learn later. Pick whichever area of the project has had least attention, and work out what needs fixing up. Once you're confident that you can deliver (if you can't, then this next step will be counterproductive), ask your tech lead if there's anything that you can put your mind to. It's likely that to start with they'll want to double-check anything you do, but, as you show that you can pre-empt and prevent problems, they'll come to trust you – and, ultimately, rely on you.

At this point, it's helpful to ask 'What is leadership?' The talent leadership programme used by the technical department at UBS sums it up as follows:

1. Notice that something needs doing

2. Work out what to do about it

3. Do it

I love this way of breaking the issue down – it both makes it seem easy ('hey! I could do that!') and gives you a clear structure for working out where you are going wrong. Is it that everything seems to be going swimmingly until you get to the end and someone points out the issues (i.e. that you're not noticing what needs doing)? Is it that you know what needs fixing but you don't know how to fix it? Or is it that you know both what is broken *and* what to do about it, but for some reason have never got around to doing it?

My problems were always at step number two. It amazed me how many times a problem would arise or be pointed out to me, and I'd realise that I had spotted it and yet let it drift away without sitting down to work out how to prevent it. On the other hand, some types of problem got dispatched swiftly. It was all to do with whether I knew what to do or not.

The fact that a person can demonstrate all the key skills of a technical lead while working as a developer makes it much easier for you, the person responsible for recruiting technical leads and ensuring that they're performing well. You don't need to take a plunge into the unknown when you appoint a new tech lead; you can simply look at how they're getting on as a developer, give them the guidance they need, and see whether they can put it into practice.

Make sure people try new things

In a lot of industries it's important to stick to the standards. Production methods are tried and tested, and any change is likely to be a regression to a previous methodology that has already been improved upon.

Software development isn't like that. The sector didn't exist before the 1950s, and it's progressing along an exponential development path. Over the 20 years that I've been working in software the industry has changed almost beyond recognition. This makes it easy to see the difference between software companies or departments that are using modern methodology and their slightly more antiquated counterparts.

Lots of companies aim to ride the wave of process innovation by reading avidly and adopting the latest fads wholeheartedly: 'We're moving over to Agile; we're going to get some trainers in, and by the end you'll all be certified Scrum masters.' The problem with this approach is that, while obviously any software leader worth their salt needs to keep on top of what everyone else in the industry is using, if you rely on trainers or mantra you have only a surface-deep understanding.

An ex-CTO of software development leaders Thoughtworks once told me that, although the concept of an Agile retrospective (a short meeting every few weeks) originated through their innovation, at many companies they have become routine box-ticking exercises, with the methodology followed but the benefits not realised. This is because many in the new generation are accustomed to learning by rote (see *There is never a substitute for using your brain* p54).

The real way to ensure process innovation is to innovate yourselves. Set your teams a goal of introducing one new practice each iteration. Make sure nothing is sacred. Want to program in teams? Fine. Want to sacrifice individual liberty and insist on the team eating lunch together every day? Fine. If it works, other teams will adopt it too. If it doesn't work, you'll only feel the pain once, and it will quietly fade away as a concept.

Technology today is a completely different world from what it was 20 years ago, and it will be completely changed again in another 20. What's more, this rapid rise of tech is changing other industries too. As a technology leader you need to be not just managing your team but driving technological business strategy through communication with the rest of the senior team. But that's a topic for a whole other book.

Create a technical career path rather than just a management one

In most jobs, when you do well at the coalface, you get promoted. And where you usually get promoted to is a post where you tell the people at the coalface what to do – i.e. management. This is an approach that I have seen used in many organisations. Based on the sound principle that you want to give more responsibility to those already delivering on the responsibility that they have, businesses look around for people who are doing well in their current job, and offer them roles as a project managers. With a few exceptions, this translates to an unwritten rule of 'the best developers get promoted to be managers'.

Now sometimes this worked really well – for example, if you take a developer who is succeeding because they are thinking about what the client wants, organising their time well, looking around for what needs doing, and picking up the unglamorous tasks that means that the software gets shipped. But not all developers make brilliant managers. On the other hand promotion of only those who have management abilities will result in a situation where less-than-stellar or even mediocre developers are put into positions of responsibility (and remunerated accordingly) above development superstars. This will not make the superstars happy. Why, when they can write you an ORM in 10 days, are they suddenly being told to write timesheets and report their status to someone who couldn't even separate their concerns?

It is important to have multiple promotion paths. With imagination you can find ways to give technical superstars real responsibility, for technical architecture and code quality, naturally, but also for the development and training of junior team members and the communication of best practice and development news. Make sure you don't lose your technical elite!

Reward using things other than money

It's tempting in today's culture to try and motivate everyone by paying them more. But that might simply not work with your technical staff. Psychologist Frederick Herzberg put forward a theory in 1959 that, while some things *do* provide positive motivation, such as recognition and responsibility, work factors such as salary, work conditions and fringe benefits are merely 'hygiene factors'. Without them, your staff will be demotivated, but in and of themselves they will not motivate people to achieve more or to do a better job.

Lots of software development offices have cool toys (fussball being one of the most common), free food and other perks. My favourite perks that I have seen include a cinema room, a music room with drum lessons and a choir, table tennis, unlimited chocolate biscuits and home-cooked lunches. But don't get confused and think this is a motivator for your team. It can more accurately be described as a hygiene factor – if your tech department doesn't have all the latest toys, they can definitely feel hard done by.

If you really want to motivate your tech team, it's important to instil passion and to fight to get them engaging work (see *Instill passion* p31). I also reward my top performers with something very exclusive: more work. It shows that they are respected and valued – that you trust them to do a good job. Plus, of course, if they are producing more, it gives you the freedom to pay them more. Win-win.

Recommended reading

These are some of the books – not all of them aimed specifically at the technology sector – that I've found useful throughout my career.

Self-leadership
The Seven Habits Of Highly Effective People, Stephen Covey
Self-Leadership And The One Minute Manager, Ken Blanchard

Leadership
How To Lead, Jo Owen

Management / People Management
The One Minute Manager, Ken Blanchard
How To Win Friends And Influence People, Dale Carnegie
5 Essential People Skills, The Carnegie Foundation
Peopleware, Timothy Lister and Tom de Marco

Technical
Clean Code, Robert C. Martin
The Clean Coder, Robert C. Martin
The Pragmatic Programmer, Dave Thomas and Andy Hunt
Working Effectively with Legacy Code, Michael C. Feathers

Networking and Sales
Never Eat Alone, Keith Ferrazzi
Give and Take, Adam Grant
To Sell Is Human, Daniel Pink

Website
You can read more about me and get in touch on my website www.zoefcunningham.com.

Thank you

Thanks to the early reviewers for their invaluable feedback. Particular thanks to Chris Harris and Nina Grunfeld's Life Clubs for getting this book out of my Dropbox and into the world.

Thank you to my first employer, Softwire, where I learnt everything in this book.

Other titles by Zoe Cunningham:

Networking Know-How!
Connecting for Succcess

Being excellent at networking is the essential skill for personal and career success, whether going after a dream job, branching out and starting a business or simply to meet new people. With a clear and practical goal-oriented approach, the book will show you how to become an amazing networker. Packed full of tried and tested techniques and secrets from some of the UK's most successful networkers, the book contains useful tools, tips and real life examples of people who have used networking to secure their dream jobs, make career moves, grow their business or even find their life partner, to those who network and connect people professionally for a living.

An Actor's Life for Me?
How to get Started in Acting

Kickstart your acting career today. Writing from hard-earned experience, and packed with interviews and top tips from contemporary actors, directors and agents, this book is your first successful step in achieving your acting ambitions. From theatre, to stage, to film and television, Zoe covers everything from initial training, methods to gain valuable experience, and on to securing an audition. The ultimate aim? To help you be selected for your first acting role!

www.ingramcontent.com/pod-product-compliance
Lightning Source LLC
Chambersburg PA
CBHW022210050726
47590CB00002B/734